THE
HORSE
BREEDING &
YOUNGSTOCK

— Julie Brega —

J.A. Allen
London

British Library Cataloguing in Publication Data
A catalogue record for this book is available from the British Library

ISBN 0-85131-647-6

Published in Great Britain in 1996 by
J. A. Allen and Company Limited,
1 Lower Grosvenor Place,
Buckingham Palace Road,
London, SW1W 0EL.

Typeset in Hong Kong by Setrite Typesetters Ltd.
Printed in Hong Kong by Dah Hua Printing Press Co. Ltd.

CONTENTS

Semen evaluation   5 5.

LIST OF ILLUSTRATIONS

LIST OF TABLES

ACKNOWLEDGEMENTS

I would like to thank the following friends for their invaluable help in the production of this book:

Debby Baker, for deciphering and typesetting the original manuscript.

Annalisa Barrelet, BVetMed, MS, Cert ESM, MRCVS, for veterinary editing and much expert advice.

Kitty Best, for the excellent illustrations.

Martin Diggle, for his patience and editing skills.

And special thanks to my family:

To my parents, John and Sheila Hollywood, who look after our children, Holly and Josh, so brilliantly, allowing me to get some work done, to my step-daughter, Zoe, for keeping the horses, office and myself organized, and finally to my husband, Bill, for his constant help and support.

INTRODUCTION

The Horse: Breeding & Youngstock is one of six books in the Progressive Series. This series forms the basis of an advanced open learning course offered by The Open College of Equestrian Studies.

The main objective of these books is to present information needed by the equestrian enthusiast in a clear and logical manner. This information is invaluable to everyone interested in horses, whether in a professional capacity as a yard manager or examination trainee, or as a private horse-owner.

This book deals with the vast subject of equine reproduction, starting with the planning stages and factors to consider when trying to decide whether or not to breed from a specific mare. Having decided to do so, the necessary considerations when selecting a stallion are dealt with.

It is the hope of everyone involved in horse breeding that each foal will inherit all of the good qualities of the sire and dam, and none of the weaknesses. Since it is the shuffling of the genes at conception which dictates the characteristics of the future offspring, the fundamentals of genetics and heredity are explained in a straightforward manner in the second chapter of the book.

This information is followed by discussion of the anatomy of the mare's and stallion's reproductive systems, and the physiology of the oestrus cycle.

In preparation for covering, both mare and stallion must be swabbed. The reasons and methods for this are explained. Since preparation also includes teasing to assess the mare's receptiveness to the stallion, the procedures for teasing and covering are described.

Artificial insemination is being used increasingly in the non-Thoroughbred breeding world. This technique is discussed, along with embryo transfer.

Methods of pregnancy diagnosis, including procedure upon diagnosis of a twin pregnancy, are dealt with next. The chapter entitled The Stages of Pregnancy explains how the foetus develops from conception to full term, and this is followed by the care and management of the in-foal mare, including the process of foaling. Care of the mare and foal in the first hours, days and weeks is then discussed, covering important aspects such as nutrition, handling, turning out and vaccination.

Pregnancy Failure and Disorders of the Reproductive System deals with causes of pregnancy failure and conditions such as contagious equine metritis, equine herpes virus and equine viral arteritis. Ailments of the foal, both non-infectious and infectious, are then discussed, in a section covering a range of problems which may affect both the newborn and older foal.

The following section covers the general management of stallions and youngstock and also deals with the first year of the foal's life, including aspects such as weaning and castration. An outline description of the initial training of the young horse, from leading and lungeing through to backing and riding, concludes the book.

1

DECIDING TO BREED

There is a worldwide surplus of unwanted horses and ponies of questionable quality — to add to this number simply because a mare has no other uses is an irresponsible error. Therefore, always consider whether you are breeding from a particular mare simply because she can do nothing else.

FACTORS TO CONSIDER IN THE MARE

There must be a very good reason for having any mare covered, so she should have all of the following qualities:

Proven ability. Has she excelled in any particular sphere — horse trials, racing, showing, etc?

Temperament. Is she free from vice, with a fairly calm and pleasant temperament?

Conformation. Is she well put together and free from any conformational defects which may affect the performance of future progeny?

Health. She should be healthy, with a normal reproductive tract — this will be ascertained by the vet upon internal examination. Another point to consider is whether any previous pregnancies or foalings have presented difficulties.

Costs

Basic costs to be incurred will include the following:

Stud fees, as agreed between the owners of both mare and stallion.

Keep fees. These will be higher if there is a foal at foot.

Transportation to and from the stud.

General management. Worming; foot care; upgrading of paddocks, fencing, etc. to make safe for a foal.

There will also be veterinary fees, which can be divided into routine, specialist treatments and emergencies.

ROUTINE VET FEES:

Vaccinations. Flu and tetanus vaccinations should be up to date, with a booster given one month before foaling. Some studs also stipulate that mares are vaccinated against the equine herpes virus rhinopneumonitis, using a killed vaccine such as Pneumabort K.

Clitoral and uterine swabbing.

Post-foaling attention — examination of the membranes and possible suturing of the vulva.

SPECIALIST VET FEES

Examination for breeding soundness.

Additional swabs.

Examination for failure to show oestrus.

Hormone treatments.

Treatment of uterine infections.

Examination to assess readiness for covering.

Pregnancy diagnosis — manual or ultrasound — which may have to be repeated more than once.

Dealing with twins.

EMERGENCY VET FEES:

Parturition problems.

Ailments affecting the mare and/or foal.

It is then necessary to consider the long-term costs of keeping a foal and young horse, bearing in mind that he may not be ridden properly until four years of age. (Although Thoroughbred foals born within the flat-racing industry may be sold as yearlings and race as two-year-olds.)

THE STALLION

A stallion should only be used at stud if he is of outstanding quality. When assessing his quality, the following points should be considered:

Achievements. He should have a good record of success in his sphere, whether eventing, racing, showing, dressage, etc. Indeed, many stallions combine a successful competitive career with their stud duties.

Conformation and movement. As with the mare, he must have no defects which may affect the future soundness of any off-spring. Moreover, his action is of great importance. A foal bred to be a racehorse must be able to gallop properly, whilst a potential dressage prospect must be even, level and straight in all gaits. Covering a mare who has good movement with a stallion who has not carries the risk that the good movement will not be passed on to the offspring. This also applies to other defects.

Temperament. How does the stallion cope with a competition environment? A steady, placid stallion *may* tone down the temperament of the offspring of a 'fizzy' type of mare.

Fertility. The stallion owner may restrict the number of mares to be covered during any season, and may refuse mares who are known to be difficult to get in foal. The stallion owner will want to advertise a high percentage of mares in-foal to his stallion to attract further clients.

√*Progeny.* Find out about the stallion's offspring and, if possible, watch them in action. It is in the interest of the stallion's owner(s) to select high quality mares for the 'book' (covering list) because the ability of the progeny substantially dictates the worth of the stallion.

Genetic potential. A foal will carry a proportionate amount of genes from each ancestor, so the family history is of great interest. Therefore, study the stallion's family tree to ascertain its quality.

Size. This is particularly important for the smaller mare about to have her first foal. It is not wise to use a relatively large stallion for covering the first time because of the increased chances of a large foal — which may present difficulties at the birth. It is, however, comparatively rare for a mare to have an oversize foal (except in some smaller breeds).

Having selected a short-list of stallions, visit the studs and see them. Many studs provide brochures about their stallions, and information on procedures and current fees.

Stud fees and associated terms

It is definitely a false economy to use a cheap stallion as a means of saving money — it is just as expensive to rear an inferior foal as it is a worthwhile one. Poor quality foals add to the glut of unwanted horses and ponies, a large percentage of which end their days prematurely at the slaughterhouse. It is therefore wise for the mare owner to choose the best stallion affordable for their purpose.

The stallion's fee normally reflects his quality or, at least, his popularity, and there are different terms upon which this fee is paid. It is always sensible to ensure that both parties are clear about these terms, and sign documents to that effect. Common arrangements are:

Straight fee. The stud fee is paid before the mare arrives at stud, or upon collection. Sometimes the fee is due at the end of the breeding season, regardless of whether or not the mare conceives. (The breeding season for Thoroughbreds ends on the 15th July.) This arrangement is often the cheapest.

Split fee. Part of the fee is due at the end of the breeding season, regardless of whether or not the mare is in foal. The balance is then due at a specified date provided the mare is in foal, or has a foal who lives for at least seven days.

No foal, no fee. The mare is tested for pregnancy and, if she is not pregnant on 1st October, the stud fee is returned. This system may also be used if the mare aborts after 1st October or the foal is born dead or does not live for a specified period.

No foal, free return. The fee is paid but, if the mare fails to produce a foal who lives to a specified age, she is entitled to return the following season to be covered by the same stallion free of charge.

2

HEREDITY – BASIC GENETICS

The science of genetics deals with the physiology of heredity – the mechanism by which characteristics are passed on from parent to offspring. For the purposes of this book, the topic of inheritance has been kept as simple as possible, providing an introduction only to this vast and complex subject.

Firstly, it must be appreciated that not all traits are inherited in what is termed simple mendelian fashion, that is to say single dominant : recessive gene inheritance. Certain traits, for example temperament, performance and fertility, are determined by the presence of a number of genes, so that several genes make up a single (polygenic) trait. It is therefore difficult to predict accurately the likelihood of the offspring inheriting the same degree of any of these characteristics from the parents. These characteristics can also be affected by environmental influences, such as management and nutrition. Fertility, in particular, is of low heritability, and is more likely to be affected by other factors.

A horse's colour is determined by arrangements of several different genes. Certain colours such as grey and black are always dominant, although the presence of other genes may modify them in some instances.

Growth/size is a highly heritable characteristic. Although nutrition does affect the rate of growth, a horse's size is genetically determined. Although defects of conformation are under genetic control, they may not always be passed on. However, to breed from stock with poor conformation in the hope that the defects will not be inherited would be risky, if not downright foolish.

In deciding whether to breed from any given mare or stallion it is necessary to consider carefully the main aims of all stock breeding. These are to maintain all desirable qualities, to improve these qualities through the generations and to eliminate undesirable qualities.

CELL STRUCTURE

In order to understand the mechanisms of basic genetics, a knowledge of cell structure is necessary. Cells are the basic units of life, of which all living things are constructed. They vary in structure depending on function or position. For example free tissue cells are spherical but, because of compaction and pressure from other cells, bone cells are spiky and skin cells are flat. Nerve cells, however, consist of long fibres.

Some of the structures of a cell can be seen through a normal light microscope. However, the detail of a cell can only be seen through a powerful electron microscope. The chief components of a cell are the cell membrane, cytoplasm and the cell nucleus.

The cell membrane

This membrane, which encloses the other components of the cell, is composed of fats and proteins, and is actively involved in the cell functions. The membrane is selectively permeable to gases, water and certain molecules, according to the requirements of the particular cell.

Cytoplasm

This substance is a highly organized living jelly, containing various organelles. These organelles include:

Mitochondria. These are minute, sausage-shaped structures packed with enzymes, responsible for the release of energy — for this reason they are often referred to as the 'powerhouses' of a cell.

Golgi bodies. These are composed of parallel, flattened bags and play a role in the secretions of glandular cells. Their functions in non-glandular cells is uncertain.

Lysosomes. These membranous compartments contain enzymes which are used to break down foreign, possibly harmful, substances in preparation for removal from the body.

Cytoplasmic inclusions. These include melanin granules, which form the pigment to give coat colouring, and also carbohydrates, proteins, lipids and water.

Centrioles. These small bodies are found near the nucleus, surrounded by condensed cytoplasm known as the centrosome. Centrioles play an important role in the reproduction of the cell.

Endoplasmic reticulum. This network of flattened, membranous, tubular cavities is filled with fluid and extends throughout the cytoplasm. Some areas are smooth while others are roughened by a covering of small particles called ribosomes.

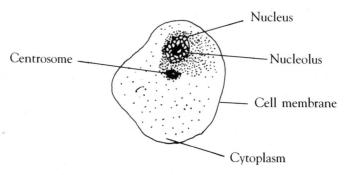

Figure 1 The components of a cell

Ribosomes. These structures are found attached to the endo-plasmic reticulum and also free-floating in the cytoplasm. They are composed of ribonucleic acid (RNA). RNA is a single helix with a phosphate/sugar backbone onto which are attached nucleic acids.

RNA is made in the nucleus by copying deoxyribonucleic acid (DNA). It migrates into the cytoplasm and is responsible for interpreting the nucleic acids on DNA and synthesising proteins as required within that cell. Information contained on the DNA within the nucleus is transported through the nuclear membrane as messenger RNA (m-RNA), to the ribosomes. Here, it forms transfer RNA (t-RNA) which influences the formation of protein characteristics of the cell.

The cell nucleus

The nucleus is a membrane-bound sac made of nucleoplasm which governs all the activities of the cytoplasm. Metabolism and growth are not possible without this 'central bank of information', so cells cannot live in the absence of a nucleus.

Within the nucleus is the nucleolus, a distinct collection of RNA responsible for the synthesis of proteins. All of the cell's genetic information is stored within the nucleus on the chromosomes in the form of genes — segments of DNA. This information passes through the nuclear membrane, which is porous and thus allows the exchange of substances between the nucleus and cytoplasm.

GENETIC INFORMATION

Within the nucleus of every cell of a horse's body, there are 64 chromosomes arranged in 32 pairs. One half of each pair of chromosomes originated from the horse's sire, the other half from the dam. Chromosomes are mainly composed of deoxyribonucleic acid (DNA). A molecule of DNA is a fine strand in the shape of a twisted rope ladder with thousands of rungs — a shape known as a double helix. The sides of the helix are made of sugar and phosphate molecules and these are held together by the 'rungs', which are made up from a pair of nucleic acid bases. These bases are adenine, thymine, cytosine

Figure 2 A DNA molecule

and guanine. Because of their chemical structures the bases only pair up as: adenine and thymine; cytosine and guanine.

The helix itself is divided into different areas known as genes. These contain the coded information necessary for the 'passing on' of various characteristics. Each characteristic is identified by a pair of genes; generally, but not always, there are two genes for each characteristic — one gene on each chromosome of each pair.

Some genes prove to be *dominant* and overshadow their *recessive* partners — the characteristic being determined by the dominant gene. In genetic science, dominant genes are indicated by capital letters, recessive genes by lower case letters. If a gene proves to be recessive that characteristic will not be present in the offspring unless, after the shuffling of genes, two recessive genes are paired up. Genes, however, are not always dominant or recessive — often they exert only a partial effect, and several genes interact to determine one single trait.

If the two genes governing a particular characteristic are the same they are said to be *homozygous*; if different, they are said to be *heterozygous*. If the genes are heterozygous one may be dominant and one recessive — in which case the characteristic will be governed by the dominant gene.

At the time of conception usually only one gamete (sperm) from the stallion will fuse with one gamete (ovum) from the mare. Each gamete contains half the number of chromosomes needed to form a new being, and thus only half of the genetic information. However, the fertilized egg, known as a zygote, begins to divide into a number of cells. Each cell now contains 32 pairs of chromosomes that is, the complete number, of which half are from the sire and half from the dam.

The characteristics of the foal will depend upon the nature of the genes for each characteristic, that is whether they are heterozygous, homozygous dominant, or homozygous recessive and which *particular* genes are dominant or recessive. It is this reorganization of the genetic information which can result in the offspring not necessarily showing the same characteristics as the parents. It can be difficult to analyse statistically the genetic expectations of a mare as, in her lifetime, she may have relatively few foals. Because of the frequency of his covering, it is easier to analyse the characteristics of the progeny of one particular stallion.

MEIOSIS

This is the process of cell division and replication within the gonads (sexual organs) of the mare and stallion, to form ovum and sperm respectively.

Providing genetic variation is an essential function of meiosis. The way in which this works is that, when the pairs of chromosomes are arranged ready for separation at the beginning of meiosis, it is a matter of chance whether the chromosomes contributed by the sperm or the ovum face a given 'pole' and any crossing over results in the exchange of information between the diverging chromosomes. This process is explained more fully in the sequence of diagrams which follows.

1) Before meiosis occurs the chromosomes are threadlike and long.

2) At the onset of meiosis they shorten, thicken and arrange themselves in pairs according to the similarities in the information contained upon the genes.

3) Each chromosome then splits lengthwise to form a pair of chromatids.

4) Crossing over takes place — the chromotids may exchange genetic material by attaching at one or more points along their length — a process known as chiasma. The chromosomes then segregate — each pair splits to opposite 'poles' of the nucleus.

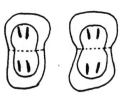

5) The cell nucleus divides into two. Each division contains 32 chromosomes and each chromosome comprises two chromatids. The point at which the chromatids are attached to each other is known as the centromere.

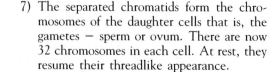

6) At this point the chromatids separate to opposite 'poles' of the nucleus. The cell constricts, so dividing into two again.

7) The separated chromatids form the chromosomes of the daughter cells that is, the gametes — sperm or ovum. There are now 32 chromosomes in each cell. At rest, they resume their threadlike appearance.

Figure 3 Meiosis — the process of gamete formation

The following diagrams illustrate how the process of meiosis can influence the exchange of information which governs different characteristics. For the purpose of hypothetical example, the genes governing the different characteristics are indicated thus:

B –	Colour (black)	–	dominant
b –	Colour (chestnut)	–	recessive
T –	Good temperament	–	dominant
t –	Bad temperament	–	recessive
Sn–	White snip to one nostril	–	dominant
sn –	No snip	–	recessive
Sh–	Sickle hocks	–	dominant
sh –	Normal hocks	–	recessive

1) Heterozygous

2) Heterozygous

3) Heterozygous

4) Heterozygous

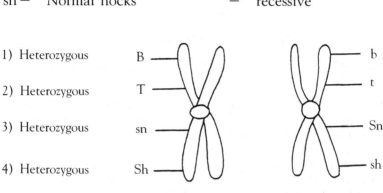

a) The chromosomes arrange themselves in pairs at the onset of meiosis. This diagram relates to stage 2 of Figure 3.

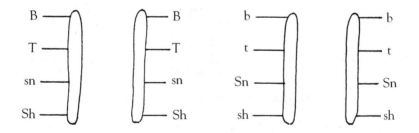

b) Each chromosome splits to form two chromatids, attached at the centromere until stage 6. This diagram relates to stage 3 of Figure 3.

The chromosomes are then well 'shuffled' – crossing over now occurs randomly, ensuring genetic variety. An example of a chromosome from this dam after crossing over may be as follows:

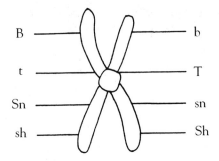

c) This diagram relates to stage 4 of Figure 3.

Figure 4 Diagrammatic representation of unions between different chromosomes — the mare.

At this stage, the pair of chromosomes migrate to opposite 'poles' of the cell nucleus, so that the chromosome shown above would be separated from its partner (not shown). After the first meiotic division (stage 5 of Figure 3) there are then only 32 chromosomes in each nucleus. At stage 6, the chromatids which make up these chromosomes separate to opposite 'poles' of the nucleus and the second division occurs. The chromatids become the chromosomes in each new gamete (stage 7).

1) Homozygous recessive

2) Heterozygous

3) Heterozygous

4) Homozygous dominant

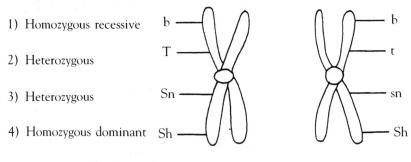

a) This diagram relates to stage 2 of Figure 3.

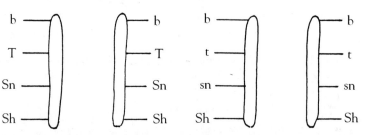

b) Each chromosome splits to form two chromatids. This diagram relates to stage 3 of Figure 3.

Having been well 'shuffled' the chromatids cross over. An example of a chromosome from this stallion after crossing over has occurred might be as follows:

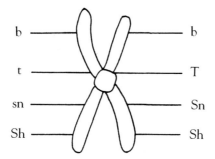

c) **This diagram relates to stage 4 of Figure 3.**

Figure 5 Diagrammatic representation of unions between different chromosomes — the stallion

At stage 6, the chromatids separate to form the new chromosomes within the gametes. At this point, using the hypothetical chromosomes of our example, we can try to deduce the possible characteristics of some of the future progeny.

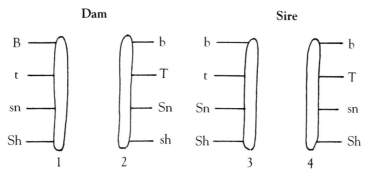

Stages 5, 6 and 7 occur as the formation of the ovum.

Figure 6 Separation of the chromatids to form new chromosomes

In Figure 6, the numbers 1 and 2 represent a sample of the genetic information contained within the chromosomes in the ovum. The numbers 3 and 4 represent a sample of the genetic information contained within the chromosomes in the sperm.

As each ovum or sperm contains only half the total number of chromosomes, they will contain only half of the genetic information necessary to produce any given characteristic. For example, one ovum will contain either number 1 or number 2, not both. The characteristics of the offspring are thus dependent on the lottery of which sperm and ovum eventually unite (see Figure 7).

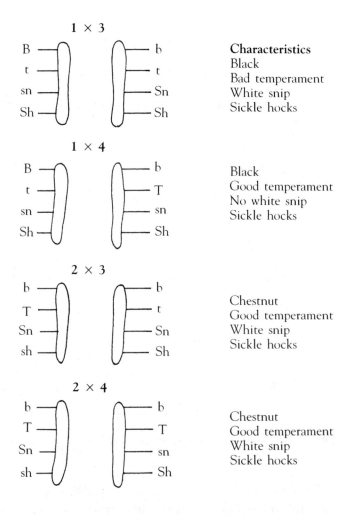

Figure 7 Representation of the effects of union between different chromosomes from sire and dam.

Sex determination

In the female, the pair of chromosomes governing sex is identical — both chromosomes are rod-shaped. They are known as X chromosomes. In the male, the two sex chromosomes differ from each other — one is a rod-shaped X chromosome and the other is a hook-shaped Y chromosome. Thus, the sex of all offspring is determined by the chromosome passed on by the sire (see Figure 8).

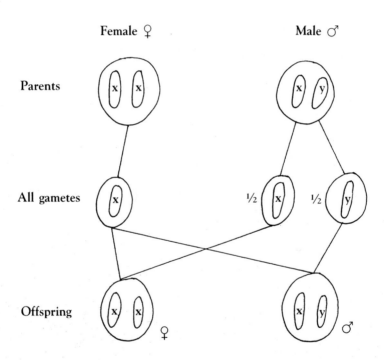

Figure 8 Chromosomal sex determination

CELL DIVISION (MITOSIS)

This is the type of cell division that takes place during growth, that is, from fertilized ovum into adult horse.

   Mitosis involves the division of cells, producing identical daughter cells containing exactly the same genetic information and number of chromosomes. One cell divides into two, two into four and so on. There are five phases in the process of mitosis. These are described in Figure 9. (For the purpose of simplicity, the 32 pairs of chromosomes are diagrammatically represented throughout Figure 9 by two pairs.)

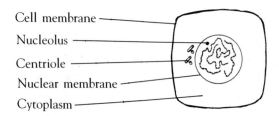

Cell membrane
Nucleolus
Centriole
Nuclear membrane
Cytoplasm

1) Interphase

The chromosomes are long and threadlike. The genetic material and cytoplasmic organelles (including centrioles) replicate to ensure that there is sufficient for all daughter cells.

2) Prophase

The chromosomes start to constrict, the nucleolus shrinks and spindle fibres (of protein) start to form. One pair of centrioles moves to the opposite side of the nucleus.

The chromosomes become shorter and fatter and replicate. They then appear as a pair of chromatids attached at the centromere. The nucleolus disappears and the nuclear membrane breaks down.

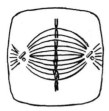 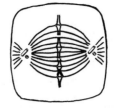

3) Metaphase

A spindle of protein fibres is now fully formed, spanning from end to end of the cell. The two ends are known as the 'poles', and the centre is the 'equator'. The chromosomes arrange themselves on the 'equator' of the spindle.

The chromatids draw apart at the centromere region. The centromeres orientate towards opposite 'poles' of the spindle.

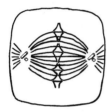 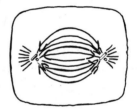

4) Anaphase

The chromatids part company and migrate to opposite 'poles'.

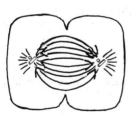 

5) Telophase

The cell membrane starts to constrict across the middle. A nucleus re-forms on both sides of the constricting membrane.

Constriction continues, the spindle degenerates and the chromosomes regain their threadlike form and return to the resting condition of interphase to prepare for the next division.

Figure 9 The phases of mitosis

3

THE REPRODUCTIVE SYSTEMS

In this chapter, we shall discuss the anatomy and functions of the mare's and stallion's reproductive systems.

THE MARE

The key functions of the mare's reproductive organs are:

1) To produce the ovum (egg), which will unite with a sperm to form an embryo.

2) To provide nutrition for the embryo.

3) To provide the perfect environment for development of the embryo.

Figure 10 shows the main features of the mare's reproductive system: let us now examine these in detail.

The vulva

The area under the tail is called the perineum. The vulva is visible externally, situated below the anus. The two external lips (vulval labia) should be opposed and vertical, forming a

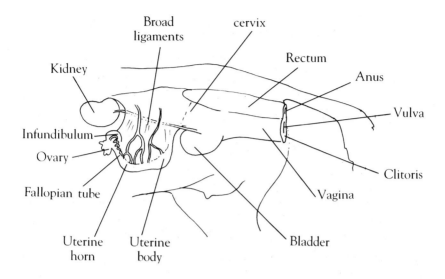

Figure 10 The mare's reproductive system

seal. If the conformation of the mare is poor (recessed anus, with a sloping vulva), a shelf below the anus is created and faecal contamination may result. Such contamination may also occur if there has been damage from previous foalings or injury from kicks, resulting in imperfect opposition of the vulval lips.

The skin covering the vulval lips is pigmented and contains many sebaceous and sweat glands. The muscular layer beneath the skin enables the vulva to lengthen or shorten according to the mare's reproductive status (whether or not she is in season, respectively).

The internal surface of the vulva is lined with a mucous membrane and is continuous with the vestibule.

The vestibule

This extends from the vulval lips to the vagina and houses the clitoris. The body of the clitoris is situated in what is known as a fossa, and contains three small cavities known as clitoral sinuses. Harmful bacteria may be present within this area, as it provides the ideal environment for microbe reproduction. For this reason, the clitoris is swabbed prior to covering.

The vulvo-vaginal constriction

At this point in maiden mares there may be a partial or complete hymen. This constriction, also known as the vestibular seal, is a natural barrier against infection from external elements, such as faeces or contaminated air. It is formed by a fold in the wall of the posterior vagina.

The vagina

This is a hollow, muscular tube which is normally completely collapsed, forming the vestibular seal. The vaginal wall, being muscular, is capable of stretching to accommodate the stallion's penis during covering and to permit the passage of the foal during labour. A mucous inner lining helps facilitate insertion of the penis and passage of the foal.

The vagina is divided into the anterior (forward) area toward the cervix and the posterior (backward) area toward the vestibule.

Protective seals

If air is drawn into the genital tract it can lead to infection and/or reduced fertility. There are three natural seals which prevent air from being drawn in. These are:

1) The vulval seal, formed by opposition of the vulval lips.

2) The vestibular seal formed by the wall of the vagina at the pubic bone.

3) The cervix.

These seals are affected by hormonal changes during oestrus.

The uterus

This hollow, muscular organ is roughly Y or T shaped, consisting of a body and two horns. It is suspended in the abdominal cavity by the broad ligaments which attach the horns of the uterus to the roof of the abdomen.

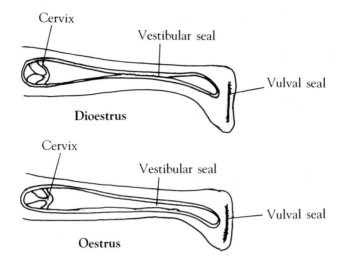

Figure 11 The three natural seals of the mare's reproductive system

When the mare is not pregnant, the uterus is relatively small and collapsed — the horns are approximately 25 cm (10 in) long, with a diameter of 1−2 cm (0.4−0.8 in) at the point where they meet the fallopian tubes. The body is usually about 20 cm (8 in) long and 10 cm (4 in) in diameter. The hind part of the body reduces to a muscular neck, known as the cervix.

The uterine wall consists of three layers:

1) An outer membranous layer (serosa) which blends with the broad ligaments.

2) A central muscular layer (myometrium), which is capable of strong contractions during labour.

3) An inner mucous membrane lining (endometrium), containing many glands and ducts.

The fallopian tubes

These are sometimes known as the uterine tubes or oviducts. One runs from the tip of each uterine horn and fans out to form a funnel structure near its ovary. Each coiled tube contains thousands of fine, flexible cilia to help propel the ovum down to the uterus. Each tube is approximately 25 cm (10 in) long. Sperm storage and subsequent fertilization occurs within the fallopian tubes.

The ovaries

The two ovaries are situated high in the abdominal cavity, just behind the kidneys. Consisting of a fibrous mass known as the stroma, they are hard and bean-shaped, but the exact size, consistency and shape is dependent upon the stage of the oestrus cycle. The sizes always vary and may range from the size of a chicken egg in younger, smaller mares, to approximately 8 × 4 × 4 cm in older, larger mares during the breeding season.

The ovaries may be felt by the vet through the wall of the rectum. There is a ligament covering each ovary which extends to the tip of the uterine horn; this is known as the utero-ovarian ligament. The whole ovary is covered by the ligament except at the ovulation fossa − a marked depression in the surface of the ovary. Each ovary abuts on the funnel-shaped entrance of its respective fallopian tube.

At birth, the ovaries contain thousands of ova, which are too small to be seen with the naked eye. No more ova will be generated during the mare's lifetime. At about two years of age the filly is sexually mature and a number of sacs develop around the ova. These sacs fill with fluid and are known as follicles. A small number of follicles increase in size during oestrus, and one (or two) will develop and enlarge prior to ovulation.

THE STALLION

The key functions of the stallion's reproductive organs are to produce sperm and to put the sperm into a position within the mare so that it may unite with an ovum to form an embryo.

The scrotum and testes

The scrotum, a sac positioned between the hind legs, houses the testes, which produce sperm and the hormone, testosterone.

The outer skin of the scrotum is thin and elastic, containing sebaceous and sweat glands. The middle layer consists of elastic, muscular tissue, while the inner layer of connective tissue is lined with a thin membrane, which allows easy movement of the testes within the scrotum.

Each testicle lies in one half of the scrotum, divided by a thin layer of tissue. The testes are roughly oval-shaped, their size being dependent on the size of the horse and the time of year. The testes of a Thoroughbred stallion are approximately $10 \times 6 \times 5$ cm, and weigh approximately 300 g (10 oz).

Each testicle is divided into lobes by sheets of fibrous tissue. Each lobe houses a network of seminiferous tubules, responsible for sperm production, and Leidig cells, which secrete testosterone.

The scrotum has the capacity to regulate the proximity of the testes to the body as a means of temperature control. Temperature control is important, as the optimum temperature for sperm production is 3 °C lower than body temperature. The blood supply is cooled as the spermatic artery divides into a capillary network before entering the testes. The optimum temperature is also maintained by the positioning of the testes

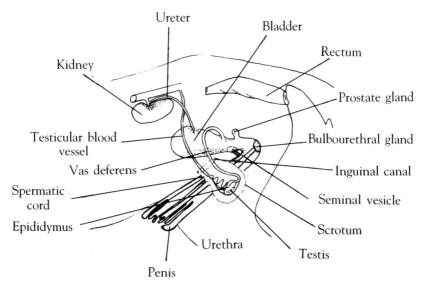

Figure 12 The stallion's reproductive system

away from the body. On very cold days, or when exercising or in danger, the testes are drawn up towards the body through the contraction of the cremaster muscle.

In the foetus, the testes are positioned close to the roof of the abdomen, near the corresponding kidney. At, or soon after birth, they descend through the inguinal canal into the scrotum. If one testicle remains undescended, the horse is referred to as a 'rig', or unilateral cryptorchid. His fertility will be affected, because the high temperature within the abdomen prevents sperm production from the retained testicle and thus leads to a lower sperm count. When neither testicle has descended, the horse is known as a bilateral cryptorchid. He may look, physically, to have been castrated, and will be sterile but, because testosterone is still produced, he will show the behavioural characteristics of an entire male.

The epididymus

This consists of a very long tube, which is tightly coiled back on itself and emanates from the anterior surface of each testicle. The enlarged front end (the head) connects to ducts emerging from the testes. The sperm formed in the testes pass into the epididymus for maturation and storage.

The vas deferens

Also called the ductus deferens, this tube is a continuation of the epididymus and transports sperm from there to the urethra. It has a diameter of approximately 1 cm and a thick, muscular wall. The vas deferens enters the abdomen in the spermatic cord through the inguinal canal — once in the abdomen, the tube dilates to accommodate a store of sperm.

The spermatic cord

This structure contains the vas deferens, the spermatic vein, lymph vessels and nerves. The blood in the spermatic artery is cooled by the surrounding network of capillaries of the spermatic vein before supplying the testicles. The cremaster muscle, adjacent to the spermatic cord, acts to vary the distance between the testes in the scrotal sac and the body, as a means of temperature regulation.

The inguinal canal

This is the channel through which the testes descend soon after birth. In some instances, a colt may have a very large inguinal canal through which a part of the intestines may pass, causing an inguinal hernia. If this is small and unnoticed, and a standing or 'open' method of castration is performed, there is a danger of some abdominal contents slipping out of the castration wound.

The urethra

This tube connects the opening of the bladder with the tip of the penis. It transports (separately) both urine and semen. Within the penis, the urethra is covered with muscle, the contractions of which add force to the passage of fluids.

Within the pelvic region there are three sets of glands — accessory glands — which empty into the urethra. These are:

1) The bulbo-urethral glands. These are positioned behind the seminal vesicles (see below), near the roots of the penis. Their secretion contributes to the seminal plasma.

2) The prostate gland. This partially surrounds the urethra and secretes a clear fluid which cleanses the urethra of bacteria and urine prior to ejaculation.

3) The seminal vesicles. A large pair of elongated sacs approximately 16–20 cm in length situated on either side of the bladder. Their secretion contributes to the gelatinous part of the seminal plasma.

The penis

The penis is housed within the sheath, the folds of which are lubricated by a natural, greasy smegma. This provides the ideal environment for microbes which cause venereal disease. The body of the penis is composed of erectile tissue, which becomes engorged with blood and fully erect when the stallion is aroused. Although this blood does not usually escape until after ejaculation a stallion *can* lose an erection without ejaculating, so loss of erection is not a reliable guide to recognizing ejaculation. During ejaculation the tip of the penis (the glans) becomes swollen further to help dilate the cervix, thus ensuring that the majority of the seminal fluid enters the uterus, and to prevent the leakage of semen from the mare. In order to avoid damage to either the stallion or the mare, the glans must return to near normal size before the stallion dismounts.

Sperm production

The complicated process by which sperm are formed is known as spermatogenesis which, in simplified terms, involves a multiplication of cells through the division of parent cells. These start off as what are known as precursor cells — spermatogonia, attached to the walls of the seminirferous tubules. They then develop into primary spermocytes, secondary spermocytes, spermatids and, finally, mature spermatozoa (in which form they are commonly referred to simply as sperm). As they develop, they migrate from the walls of the seminiferous tubules and each develops a tail, which only becomes functional once the spermatozoa are mature. Sperm develop within the epididymus. The average time taken for spermatogonia to mature into spermatozoa is between sixty and seventy days. This process occurs in waves to ensure a constant supply of mature sperm.

There are three distinct areas of a sperm cell: the head, which contains within the nucleus the genetic material which ultimately fuses with the nucleus of the ovum; the midpiece, which is made up of a helix of mitochondria (energy-producing cells); and the tail, made up of muscle fibrils which, using energy produced by the midpiece, propels the sperm. Sperm, however, are immotile (incapable of spontaneous movement) until mixed with fluids from the accessory glands at ejaculation to form the substance known as semen.

Semen

The fluid emitted during coitus is referred to as the ejaculate. This is a whitish, gelatinous fluid, consisting of seminal plasma and sperm.

Seminal plasma is formed by the accessory glands and provides the medium containing the elements necessary for sperm survival. Semen is ejaculated in three distinct fractions:

First, pre-sperm fraction, which consists of approximately 10 ml of bulbourethal gland secretion, and contains no sperm.

Second, sperm-rich fraction. Normally 40−80 ml containing 80−90 per cent of the sperm.

Third, post-sperm (gel) fraction. Secreted from the seminal vesicles, the volume varies from 0−80 ml dependent on season (lower out of breeding season), age, breed and previous use.

Semen samples may be taken so that the quality and concentration of sperm may be evaluated through specialized testing. These tests determine, among other things, the number and concentration of live and dead sperm, normal and abnormal.

4

THE OESTRUS
CYCLE

A mare comes into season as a result of hormonal changes. The breeding season normally begins around February/March and ends around September/October, with the oestrus cycle occurring regularly throughout this time — all mares will vary individually. During the period between October and February there is no ovarian activity — this is known as anoestrus. The majority of mares are sexually mature at around two years of age, although some early foals may ovulate as yearlings, and some Thoroughbreds do not mature until around four years of age.

In early spring, the rise in temperature, extended daylight hours and improvement in the quality of the grass stimulate the anterior pituitary gland to produce follicle stimulating hormone (FSH). This hormone stimulates the ovaries and so the oestrus cycle begins (see Table 1.)

First, one of the thousands of follicles present grows rapidly, rather like a hard cyst on the surface of the ovary. It may measure approximately 60 mm when ripe. This follicle has short-lived powers of hormone production — its innermost layers produce oestrogen, which causes increased mucus production and circulation within the vagina, cervix and uterus. The oestrogen triggers off the visible signs of oestrus, signalling that the mare is 'in season', sometimes referred to as being 'on heat'. This is the time during which the mare will accept the stallion. It normally lasts between four and seven days, although this is quite variable from mare to mare.

The pituitary gland now produces Luteinizing hormone (LH) which encourages further growth and maturity of the follicle. Upon reaching maturity the follicle softens and ruptures. Its contents escape into the fallopian tube, via a specialized indent of the ovary known as the ovulation fossa. This is the process of ovulation, which usually occurs approximately four days after the mare comes into season — sometimes this is delayed in young mares, or when there is a hormone deficiency.

The empty follicle soon fills with blood and is now termed a corpus haemorrhagicum. It can be felt through rectal examination, and palpation of the blood-filled follicle causes the mare to flinch with discomfort. The innermost layer of cells within the follicle grow rapidly — they replace the blood and form a soft, cyst-like yellow body in the ovulation fossa. This yellow body is known as the corpus luteum, a temporary but very important gland, which produces large amounts of progesterone for the next fourteen days before being resorbed.

The progesterone reverses the effects of the oestrogen, so ending oestrus and thus the mare becomes disinterested in mating. Whilst inhibited by large quantities of progesterone, the pituitary gland ceases to produce the sex hormones. If the ovum has been fertilized, the corpus luteum will continue to produce progesterone, which stimulates preparation of the uterus for pregnancy whilst further suppressing oestrus.

If the ovum has not been fertilized the uterus produces the hormone prostaglandin, which destroys the corpus luteum. The corpus luteum will begin to be resorbed (luteolysis) and the levels of progesterone in the bloodstream will decrease, allowing the pituitary gland to resume production of the sex hormones so that the cycle will start again.

The usual length of the whole oestrus cycle, that is the period between the first day of heat in one cycle and the first day of heat in the next, is twenty-two days. This can, of course, vary from month to month and from mare to mare. The observant owner must learn what is normal for a particular mare.

Table 1. The Oestrus Cycle, Showing Hormonal Activity

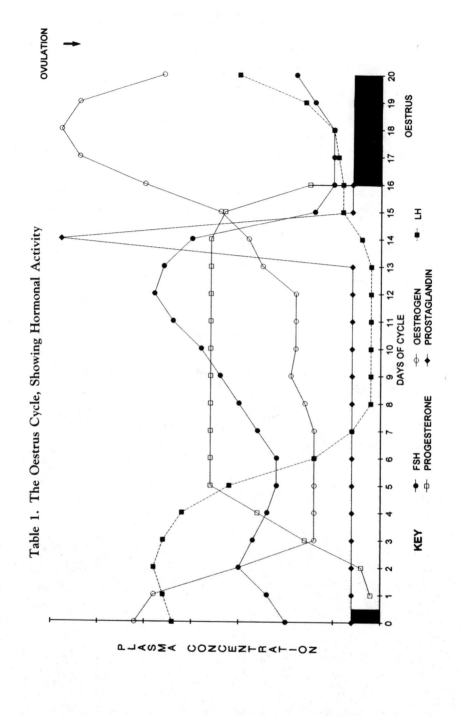

OVULATION

OESTRUS

DAYS OF CYCLE

PLASMA CONCENTRATION

KEY

FSH ●   OESTROGEN ⊕   LH ■

PROGESTERONE ⊡   PROSTAGLANDIN ◆

Table 2. The Oestrus Cycle, Showing Physiological Changes

	Ovaries	Uterus	Cervix	Vagina
Anoestrus	Small, hard, bean-shaped, with very small follicles.	Thin walled and flaccid.	Pale, dry/tacky and closed.	Pale and dry/tacky.
Transitional period	Become larger and softer. Waves of follicular growth. Often feel like 'bunch of grapes'.	Flaccid, beginning to develop some tone.	Soft.	Dependent on stage of follicular waves, either moist or dry.
Oestrus	Follicles well developed and soft.	Becomes thickened, swollen and flaccid.	Moist, pink and swollen.	Progressively more moist and pink.
Ovulation	Just before ovulation one (or two) follicle(s) become(s) very soft. Follicular fluid and ovum released. Ovary tender.	As above.	Completely relaxed.	As above.
Post-ovulation	Ovulation fossa fills with blood – corpus haemorrhagicum (CH).	As above.	Cervix starts to tighten.	As above.
Dioestrus	Progesterone produced. CL begins to develop from CH. Ovary reduces in size.	Uterine wall has 'tone' and feels turgid.	Begins to tighten. Becomes pale and dry.	Reduced mucus, becomes pale and dry.

Dioestrus, sometimes referred to as interoestrus, is the period between the last day of one heat and the first day of the next; this is usually fourteen to sixteen days.

THE FOAL HEAT

Around five to nine days after foaling the mare may come into season. This is known as the 'foal heat'. The advantages of covering the mare during the foal heat are:

1) It is normally easy to identify, as the exact date of foaling is known.

2) The mare will still be at stud, thus reducing time and costs involved with transport.

3) For a late foaler, it may be the last chance of conception in that season.

4) It is very useful for mares known to have erratic cycles after the foal heat as it avoids the confusion of calculating the date of ovulation in subsequent heats.

The disadvantages of using the foal heat (which usually outweigh the advantages) are:

1) Damage to the uterus may make it undesirable to cover the mare — if there is a problem with the endometrium, covering may cause permanent damage, possibly affecting future fertility.

2) There is a lower conception rate, as the endometrium has not had long enough to recover fully following foaling, and may not be ready for implantation.

3) There is a higher death rate associated with foal heat coverings.

4) It may not be desirable if the mare has foaled early, since covering would result in an even earlier foal next year.

The foal heat should never be used if the mare suffered dystocia (a problem foaling) or retained placenta.

RECOGNIZING OESTRUS

All mares vary in the signs they show during oestrus. The classic signs usually include the following:

The mare will 'squat' slightly, straddling her hind legs, raising the tail and everting her clitoris ('winking').

Relaxation of perineal/vulval tissues — lengthening of vulva and relaxation of cervix. This may be accompanied by the showing of a clear mucous discharge and/or urination.

Some bad-tempered mares become less aggressive, whilst normally placid mares may become more sensitive, particularly when being groomed, rugged up, etc. Some mares 'lean' on people.

These signs may be apparent generally, for example when the mare is out in the field, or a teaser may be used to aid diagnosis. Sometimes it is difficult to ascertain whether a mare is in season or not, in which case it may be necessary for the vet to examine the mare by rectal palpation.

Examination of the mare's genital system

Throughout the oestrus cycle, the organs of the genital tract undergo many changes as previously described and listed in Table 2. The external changes are visible to the layman, but any internal changes are only detectable through examination by the vet. It may be necessary to identify changes in order to ascertain the stage of the oestrus cycle, to determine whether or not the mare is pregnant, or to help identify a disorder of the genital tract. The vet will carry out such examination as is necessary either manually, or through the use of ultrasound.

Manual examination

Rectal palpation. The vet inserts a well lubricated, gloved hand into the rectum in order to feel the ovaries, cervix and uterus.

Vaginal examination. The vet inserts a hand into the vagina, in order to palpate the vagina and cervix. Sometimes a speculum is used to facilitate visual examination. This is a hollow tube used with a light source which, once inserted, allows the vet to visualize the vagina and cervix. A sterile swab can also be inserted through the speculum to obtain a bacteriological swab or endometrial smear. Samples of the uterine wall (uterine biopsy) may be taken with a special basket-jawed biopsy instrument.

Ultrasound examination

Ultrasound examination uses rapid pulses of high-frequency sound waves emitted and received by a transducer, which is inserted into the rectum. These sound waves are reflected by dense tissue such as bone, and the echo received by the transducer is then displayed on a screen. The dense tissue shows up as white, while liquids which transmit the waves appear black. This method of examination can often give a more accurate insight into the state of the reproductive tract — for example, follicles on the ovary can be seen and can be differentiated from corpus haemorrhagicum or corpus luteum.

5

THE COVERING PROCESS

In order to maximize the chances of successful covering, thorough preparation is necessary. The first procedure undertaken is swabbing.

SWABBING

Swabbing is a quick and efficient method of assessing the presence of bacterial contamination within the reproductive tract of both mare and stallion. It is, indeed, an essential practice for the following reasons:

1) Mucosal secretions may be tested for the presence of pathogenic (disease causing) organisms.

2) It prevents the transmission of venereal bacteria from the stallion's genitalia to that of the mare, or vice versa.

3) It assists in the detection of endometritis, thus helping to optimize fertility.

Swabs are therefore taken as part of the preventative routine at all studs in accordance with the *Common Codes of Practice for the Control of Contagious Equine Metritis and other Equine Bacterial Venereal Diseases and Equine Viral Arteritis*. The Codes are published by the Horserace Betting Levy Board and apply to the breeding season in France, Germany, Italy, Ireland and the UK. The Codes are reviewed and re-published annually and are available from the Thoroughbred Breeders' Association, Stanstead House, The Avenue, Newmarket, Suffolk.

Regarding contagious equine metritis (CEM), mares are defined as being either 'high risk' or 'low risk'. The following are defined as 'high risk':

1) All mares from whom the contagious equine metritis organism (CEMO) has been isolated since 1987, and any mare from which CEMO was isolated prior to 1987 and who has not since been covered.

2) Any mare covered by a stallion who transmitted CEM the previous year.

3) All mares arriving from countries other than France, Germany, Ireland, Italy and the UK, if covered by stallions resident outside these countries in the previous year.

4) Maiden mares arriving from countries other than France, Germany, Ireland, Italy or the UK.

All mares not defined as 'high risk' are designated 'low risk', and are subject to less stringent testing than the 'high risk' mares.

Should a mare be covered in France, Germany, Ireland, Italy or the UK, return to another country and subsequently come back to one of the participating countries, she will be regarded as 'low risk' provided she has *not been covered* in a 'high risk' country.

Types of swabbing

Clitoral swabbing

This is carried out to detect the presence of venereal disease pathogens. The mare will need to be well restrained and the tail bandaged and/or held out of the way. If it is very contaminated with faeces, the vulval area is wiped with a dry paper towel. No other cleansing should be carried out. The clitoral fossa and central clitoral sinus are swabbed and the swabs are immediately placed in a special transport medium and sent quickly to the laboratory. Special narrow-tipped swabs are used so they can penetrate into the small sinus. The swabs are cultured for a minimum of four days under special atmospheric conditions to test for CEMO.

Endometrial (uterine) swabbing

This is an essential practice when diagnosing endometritis (inflammation of the uterus) and is normally carried out in early oestrus before covering. This test helps to detect the presence of pathogens within the uterus. As there are bacteria present around the vulval area, it must be ensured that the swab only collects bacteria from the uterus. The vulva is cleaned thoroughly with warm, clean water and the tail bandaged. The swab is passed via a speculum through the cervix into the uterus. Once withdrawn, the swab is placed in the transport medium immediately and either taken or sent to the laboratory. If the swab is to be posted it should be done well in advance of the weekend to ensure arrival on a working day and prevent delay.

This form of test is not carried out if there is a possibility that the mare is pregnant.

Endometrial cytology (smear test)

A smear sample is collected in the same way as an endometrial (uterine) swab. The swab is rolled onto a microscope slide to collect the cells. The sample is fixed and sent to the laboratory for cytological examination (examination of cells).

In a normal 'clean' mare, endometrial lining cells are found. If there is inflammation present pus cells (PMN's) will be found on the smear. The smear test gives a more accurate indication of the presence of inflammation than does the bacteriological swab alone. The two tests should be used in conjunction — the bacteriological swab is essential to screen for venereal pathogens.

Endometrial biopsy

This may be used as a diagnostic aid. A sample of tissue is taken from the lining of the uterus with the basket-jawed biopsy instrument. The sample is then sent to the laboratory in a preservative solution for testing. This can be useful in helping to determine why a mare fails to conceive, or maintain a pregnancy.

Cervical swabbing

These swabs are easier to obtain than uterine swabs and may be collected at any time during the oestrus cycle. The process is similar to uterine swabbing but the swab is not passed through the cervix, instead secretions from the cervix itself are absorbed for testing. Unless the mare is in oestrus, this method does not give reliable information about the uterine environment.

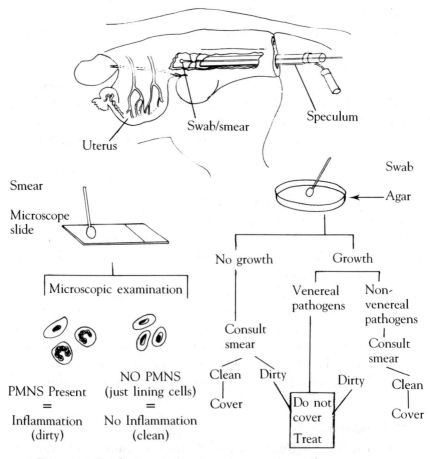

Figure 13 Producing, and acting upon, swabs and smear tests

Swabbing the stallion

The surface of the penis is swabbed to test for the presence of venereal disease pathogens, which thrive in the smegma found around the sheath area. Routine screening of all stallions is desirable as they are often only carriers of disease — they do not normally show any clinical signs of infection.

The screening requirements are usually related to the value of the animals involved — it is usual for a Thoroughbred stallion to have two sets of swabs taken annually, between 1st January and the start of the breeding season. *The Code of Practice* recommends swabbing the urethra, urethral fossa and sheath and testing a collection of pre-ejaculatory fluid.

Table 3. Summary of Swabbing Procedures Carried Out After 1st January in any Year

	Type of swab	When	Where
'LOW RISK'	Clitoral	Prior to arrival at stud OR	Home (by agreement with stud manager)
MARES		On arrival	Stud
	Endometrial	During oestrus prior to covering Repeated in subsequent oestrus if not in foal	Stud
'HIGH RISK'	Clitoral	First: before arrival at stud	Home
MARES		Second: after arrival	Stud
		Third: during oestrus	Stud
	Endometrial	During oestrus Repeated in subsequent oestrus if not in foal	Stud
STALLION	Urethra, urethral fossa, penile sheath, samples of pre-ejaculatory fluid	Before start of season, two sets taken at seven-day intervals	Home/Stud

CERTIFICATION

The mare certificate

This is completed by the mare's owner and lodged with the prospective stallion's owner or stud manager before the mare is sent to the stud. The certificate should include the following information:

Name of mare.

Passport number (if available).

Owner's name and address.

Details of studs visited, previous coverings and outcomes.

Additional information, including results of positive bacteriological examinations for CEMO, Klebsiella pneumoniae (including capsule type − see Laboratory Certificate), and Pseudomonas aeruginosa (infectious venereal bacteria).

The stallion certificate

This is a certificate of examination of the stallion, to be completed by a veterinary surgeon. It should contain the following information:

Name of stallion.

Passport number (if available).

The two dates upon which swabs were obtained from the penile sheath, urethra, urethral fossa and a sample of pre-ejaculatory fluid.

Name of the Designated Laboratory.

Confirmation of negative results.

Name, address and signature of the veterinary surgeon responsible.

The laboratory certificate

This is issued only by a 'Designated Laboratory' — one whose name is published in the *Veterinary Record* by the Horserace Betting Levy Board. The certificate should include the following information:

Name of stallion or mare the swabs were taken from (as labelled on transport medium container).

Sites swabs taken from.

Who submitted swabs.

Date that swabs were bacteriologically examined.

Name and qualifications of person responsible for examination.

Name of laboratory.

Confirmation that CEMO, Klebsiella pneumoniae or Pseudomonas aeruginosa was or was not isolated. (If Klebsiella pneumoniae was isolated, the capsule type must be stated. Only types K1, 2 and 5 are venereal pathogens.)

Covering should not take place until the relevant certificates, confirming negative results, are received.

TEASING

This procedure is also known as 'trying'. The object of teasing a mare is to see whether she is in oestrus and ready to accept the stallion. As mentioned earlier, mares vary in their indications which show oestrus — some mares show the signs of coming into season but are not ready to accept the stallion.

Ovulation occurs at the end of behavioural oestrus and the ideal time for covering is twenty-four hours before ovulation. The vet may be able to feel the stage of oestrus through manual palpation, but the best way of testing whether the mare will accept the stallion is to tease, or try, her.

The exact method of teasing may vary from stud to stud, according to their tried and tested methods and experiences. The teaser *may* be the stallion who is to cover the mare, but this is not normally the case, especially when the stallion is of high value. Instead, to avoid any risk of injury to the stallion, another animal is employed. Sometimes, a teaser stallion is kept specifically for that purpose. This can be a relatively inexpensive method if a pony stallion is used. He may be allowed to actually cover the occasional mare (although not those sent for the services of the main stallion!) to keep his interest in his job. Alternatively, a 'riggy' gelding who displays the normal characteristics of a stallion may be used.

All teasers, whether stallions or geldings, must undergo the same swabbing procedures as the main stallion. Mares who are covered by the teaser must also be fully swabbed, to avoid the risk of venereal disease being introduced to the stud. Failure to observe such precautions often constitutes the highest risk of disease spreading to studs.

The teasing board

This is a very strong, well built barrier which is high and long enough to prevent the mare from striking out and/or kicking the teaser. It must be solidly constructed and preferably covered in heavy-duty rubber. The top must be rounded, smooth and also covered in rubber to reduce the risk of injury in the event of the stallion putting a foreleg over the top.

An indoor board is useful when covering early in the year, as at Thoroughbred studs. Sometimes, an enclosure is used for the teaser so that he may be unrestrained. In such cases the sides must be high enough to deter him from jumping out — which may mean that a rail is needed over the teasing board itself.

Some teaser boards are built into the fence line of a paddock — the mares who are 'ready' come to the board when the stallion is led up to it.

It is not satisfactory to tease over a stable door as the mare may injure herself on the bolts. The stallion may also hit his head on the top of the door frame.

Teasing procedure

The teasing procedure is usually time-consuming and labour-intensive. There is also risk to the handlers, especially if the stallion becomes overexcited and lacks manners. It is, however, an essential part of good stud management. Handlers are advised to wear gloves and protective headwear. Each yard has its own guidelines for staff to follow.

The mare is led to the board wearing a bridle. The teaser, wearing his bridle, is led to stand the other side of the board by his handlers. If the mare is being tried in the foal heat, her foal will need to be restrained nearby and in sight to avoid the risk of injury. Most mares become distressed if separated from their foal.

At first, the teaser and mare may sniff nose to nose. The mare's initial response may be misleading — she may show signs of disinterest; squealing, striking out and kicking (at which point the handlers must beware), but she may change her attitude after a few more minutes. The mare should be lined up alongside the teasing board and the teaser should be allowed to sniff her body, working towards her vulva. If the mare is ready she will stand still and show the signs of oestrus quite clearly. Problems may occur if the teaser becomes overexcited or aggressive — it may require two experienced handlers to constrain him.

If the teaser is led to the field, the interested mares will normally come up to a teasing board in the fence line in response to his call. However, shy or submissive mares may not come to the fence, and may need to be teased in hand.

COVERING PROCEDURE

Studs vary in their policies regarding the handling and restraint of the horses during covering. The pasture method, whereby the stallion runs loose with his mares, is the most natural and least labour-intensive method, but it is not usually practised with valuable stallions because, although most stallions use their instincts to avoid being kicked, there is still the risk of injury. The other disadvantages of this system are:

The stallion may show a preference for some mares and ignore others, even though they are very well in season.

The stallion may cover fewer mares because of the frequency of covering, which is very energy-sapping.

It is difficult to be accurate about dates of conception.

For these reasons, and because covering in hand requires more human intervention, it is this second method that we will consider in detail.

Preparation of the mare

The anus, dock and vulva should be cleaned with warm water and then dried with a disposable paper towel. When washing the perineal area, gloves should be worn and cotton wool swabs used. A new swab should be used for each wipe, and then be disposed of.

(Note that the repeated use of soaps and antiseptics for washing off disrupts the normal skin grease and microflora, and is to be avoided at all times except under the specific instructions of a vet. The use of antiseptics also encourages overgrowth by resistant bacteria, such as Klebsiella and Pseudomonas.)

The tail is bandaged to prevent hairs from lying over the vulva as the stallion tries to insert his penis. Felt, canvas or leather covering boots are fastened onto the mare's hind feet to help soften the blow from any kick. (Occasionally, mares take a dislike to the boots and try to kick them off.) If the stallion is known to be over-vigorous with his forefeet or teeth, the mare may need to wear a protective leather pad over her neck and shoulders.

Preparation of the stallion

Most stallions wear a specific bridle when covering. There may be a long rope or, in the case of an unpredictable or difficult stallion, a pole attached to the bridle.

The stallion should have his genital area including his penis washed, as well as his belly and the insides of his hind legs. Warm water can be used; the penis should be washed when erect.

Methods of restraint

Mares are restrained in order to reduce the risk of injury to a valuable stallion but restraint should, wherever possible, be kept to a minimum. An experienced mare who is well in season should only need to wear a bridle and the protective boots. Other means of restraint are:

1) A twitch (the rope type, not a metal one). This may be necessary on a nervous mare, or one whose temperament is unknown. The handler must be able to stand well clear in case the mare strikes out.

2) Lifting and holding one foreleg with either a rope or a leather strap. This must have a quick-release mechanism so that the leg can be freed once the stallion has mounted, or if the mare stumbles.

3) Hobbles — these fasten around the hind legs and are attached to a neckstrap by very strong cords. The hobbles are fitted so that they only come into action if the mare strikes out, and they must have a quick-release mechanism in case either the mare or stallion becomes entangled. Hobbles are an extreme method of restraint, not in common use in the UK.

If a mare is being covered with a foal at foot, another handler will be needed to keep the foal safely in front of the mare.

Covering

Stallions normally associate being led to the covering yard with the procedure to follow and become aroused accordingly. This excitement may be shown as vocalization, rearing and bucking — which may make the stallion difficult to handle. His behaviour will depend upon his early training, daily handling, temperament and the ability and experience of the handler.

The stallion is normally led to the mare's side, not behind her, as she may try to kick. If a tall mare is to be covered by a smaller stallion, it may be necessary to stand her in a hollow.

The stallion will then start to tease the mare by vocalizing, licking, nuzzling and biting. The Flehman posture may be exhibited: the stallion curling back his top lip and extending his head and neck upwards and forwards. It is thought that by curling back his top lip the stallion is able to smell the mare's scent more effectively.

The stallion should by now have achieved an erection − this may occur very quickly or after a very long wait − all stallions differ in the amount of time taken. (In the case of the stallion requiring a long period in which to achieve an erection, it may be safer to stand him and the mare either side of the teasing board until he is ready to mount.)

Mounting

Whilst young, a stallion should be trained to mount at the correct time − that is, once he has achieved a full erection. An inexperienced stallion may try to mount the mare sideways, in which case he must be encouraged to dismount, and the mare must be turned for him. (Young stallions should ideally be taught using experienced mares of the correct height.)

When the stallion is mounting, all handlers should stand on the same side to avoid being kicked. In the event of a problem, the mare and stallion can be pulled towards the handlers.

Intromission

Most stallions are able to enter the mare unaided. The mare's tail may have to be held to one side to help the stallion, but he may object to any handling of his penis, which can disrupt the procedure and affect ejaculation.

Upon intromission the mare may take a step forward to help retain her balance. It is important that she is kept straight and is not allowed to take too many forward steps. It must be checked that the stallion does not try to insert his penis into the mare's rectum. Once the stallion has successfully gained intromission, he will start to thrust.

The desired stallion may be a great distance away — the semen can be quickly transported, saving time and money travelling.

The mare may be injured or unable to travel for some reason; for example, she may be a bad traveller, or fail to settle in new surroundings.

The mare may be susceptible to uterine infection — AI can help reduce the transmission of venereal and other diseases.

Each ejaculate collected may be diluted in a special solution of nutrients and antibiotics known as extender. This solution may then be divided into several portions, enabling more mares to be serviced, and may help the fertility rate of a sub-fertile stallion.

The semen can be checked for quality more frequently.

If specially frozen in liquid nitrogen, semen may be stored for use after the death of a stallion, or after his retirement from active stud duties.

At the time of the desired covering, mare or stallion may be competing.

AI PROCEDURE

Having chosen a suitable stallion, the mare owner will complete a Nomination Agreement for Artificial Insemination and agree to the terms and conditions as specified by the individual stud.

Veterinary examination prior to AI

Once the stud accepts a mare, the vet must examine her and forward a certificate to the stud stating that:

Upon internal examination, no reason was found why she should not be fertile and able to carry a foal.

The mare has no congenital abnormalities or any known hereditary diseases.

A cervical swab, which showed no sign of the presence of pathogenic bacteria indicative of uterine infection, was taken from the mare during oestrus.

Detecting oestrus

The mare's owner will need to be very observant, looking out for the signs of oestrus. Records must be kept detailing when she came into oestrus, and for how long. If she does not appear to be coming into season the vet should examine her.

Once she shows signs of oestrus, the vet must be called to palpate the ovaries in order to assess when she is likely to ovulate. The optimum time to inseminate is from twelve hours before to eighteen hours after ovulation. The vet may give an injection of luteinizing hormone to hasten the time of ovulation, enabling a more accurate prediction of when this might occur. The stud should be given provisional notice at this time.

When the vet has decided upon the optimum insemination date, the stud must again be notified. The stud will need approximately twenty-four hours notice so that the semen can be collected and dispatched.

Collecting semen

This is normally undertaken by experienced stud grooms. The stallion may be encouraged to mount an in-season mare — his penis is then guided into an artificial vagina (AV). The AV is the receptacle normally used for the collection of semen as it provides an environment very similar to a real mare's vagina.

There are several different models of AV available but they are all of similar design. They consist of a tough yet lightweight outer case, which contains a small tap and has a handle attached. The casing has a rubber liner. The space between the outer casing and rubber lining is filled with sufficient warm water to maintain the correct temperature, 40–44 °C (104–111.2 °F) and to exert pressure within the lumen of the AV, thereby mimicking a natural vagina. The temperature should never exceed 50 °C, as the sperm could be damaged and it might cause discomfort to the stallion, possibly affecting his willingness to use the AV on future occasions.

The AV can also be lined with a disposable liner bag, lubricated with sterile obstetric lubricant, to reduce the risk of infections being transferred between stallions. However, some stallions dislike the texture of these liners and will not ejaculate when they are used.

The semen may be collected in the bottom of the liner bag, or via a filter in the bottom of the liner leading to a collection vessel. Prior to use, the collection vessel must be warmed to the same temperature as the lumen of the AV. The AV and collection vessel can then be encased in a protective jacket to maintain the correct temperature and protect the semen during and after collection, since sperm are damaged by ultraviolet light and sudden changes of temperature.

The stallion can be trained to ejaculate into the AV using a mare in oestrus and allowing him to mount her. Mares used for this procedure are referred to as 'jump' mares and need to be of a calm disposition. However, because of the risks involved it is safer and more convenient to train the stallion to use a dummy or 'phantom' mare. He may be introduced to the dummy in the presence of an oestrus mare. (Stallions with a low libido may never be keen on using the dummy mare.)

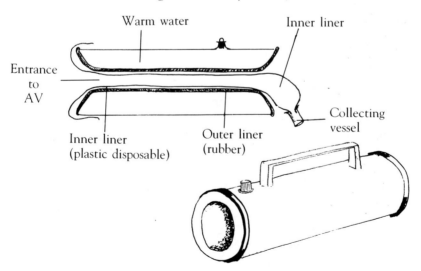

Figure 14 An artificial vagina − cross section and outer view

Once the semen is collected, the gel is removed and the semen is mixed with extender fluid at a ratio dependent upon the concentration and motility of normal sperm within the sample (see Evaluation of Semen). The extender contains protective, energy-rich substances which provide the ideal medium for storage whilst promoting the survival chances of the sperm. The semen is then placed immediately in the special 'Equitainer' bucket which cools it slowly and then maintains it at 4°C. There are two types of 'Equitainer' buckets — the 'Equitainer 40' maintains the optimum temperature for forty hours after collection from the stallion, while the 'Equitainer 70' maintains the correct refrigerated conditions for up to seventy hours.

If the semen is to be used for mares nearby very quickly, it may be inseminated 'raw' — without going into the extender solution.

Fertility expectations of chilled semen

The fertility expectations of the chilled semen are as follows:

Up to 24 hours	Excellent
24–48 hours	Good
48–60 hours	Fair
60–72 hours	Poor

These expectations are dependent on the stallion and the ejaculate — they are not constants, indeed not all stallions produce semen which tolerates storage. Furthermore, the fertility of any semen will be adversely affected by one, or a combination, of the following factors:

Exposure to sunlight.

Temperature shock.

Careless AI procedure.

Careless management of the mare.

Exposure to spermicidal substances which include:
  detergents and tap water;
  any product containing preservatives;
  disposable syringes, especially those with rubber-ended plungers;
  some brands of disposable gloves;
  most lubricants;
  air.

Evaluation of semen

Before the semen is used for AI it has to be evaluated to ascertain the concentration of normal motile sperm in the sample. The semen must be kept at $38-40\,°C$ ($100.5-104\,°F$) whilst being examined, so motility is assessed immediately on a pre-warmed slide under a microscope.

The motility of the sperm is graded on a scale $0-5$.

0 = Immotile, therefore very poor.

1 = Stationary or weak rotary movement — poor.

2 = Fewer than 50 per cent are progressively motile and there are no waves or currents.

3 = Progressively rapid movement and slow waves, indicating that approximately 50 per cent are progressively motile.

4 = Vigorous, progressive movement. Rapid waves indicate that approximately 60 per cent are progressively motile — very good.

5 = Very vigorous forward movement, strong, rapid currents indicating that more than 60 per cent are progressively motile — excellent. ($60-80$ per cent is the maximum range.)

To be suitable for use in AI the semen must contain at least 40 per cent progressively motile sperm (moving in relatively straight lines) that is, grade $2-3$ and above.

Next, a diluted fixed (killed) sample of the semen is examined under the microscope to assess the number of sperm showing abnormalities. This number is expressed as a percentage of the total number of sperm. The ratio of live : dead sperm is also measured. Semen must contain at least 65 per cent normal sperm and have a minimum ratio of $6 \cdot 5 : 3 \cdot 5$ live : dead sperm.

A special instrument is used to calculate the concentration of sperm within a sample. Once it is confirmed that the sample is suitable for use, it can be used in one of four ways:

1) Used immediately, undiluted, to inseminate one or two mares.

2) Diluted and used immediately to inseminate several mares.

3) Diluted and refrigerated to be used within sixty hours.

4) Diluted and frozen for use at a later date.

Frozen semen

Using frozen semen has both advantages and disadvantages.

ADVANTAGES

Semen can be preserved almost indefinitely.

The horse can be used for stud purposes whilst continuing a full time competition career.

Semen can be shipped long distances.

Semen can be kept so that it is ready for use when the mare is at optimal mating time with less complex co-ordination.

DISADVANTAGES

Pregnancy rates are generally much lower than with fresh/ chilled semen.

Much higher levels of knowledge and skill are required for successful fertilization.

Mares must be inseminated within six hours of ovulation. This requires a high level of veterinary input and, consequently, expense.

Because of the inherently lower pregnancy rates, it is not advisable to use frozen semen for old mares or those with suspect fertility.

Dispatch of frozen semen

The stud will normally measure the density of sperm in a sample of semen from each ejaculate, and calculate the quantity of extended semen to be used in each insemination. Typically, mares are inseminated with 250−500 million progressively motile sperm. It is usual for between 20 ml and 80 ml of semen to be supplied in each consignment − this is enough for two inseminations. The stud will probably keep a control sample for a few days.

The semen, in its 'Equitainer' bucket, will be transported either by private arrangement or by British Rail Red Star or Datapost. Consideration must be taken for weekends and Bank Holidays. Once the 'Equitainer' bucket arrives it must not be opened until the vet is ready to inseminate, as the cooling system will be disturbed. Before use, the vet will double check that the semen sample has not been adversely affected by transportation: any alterations of temperature caused by delay will adversely affect the quality of the semen.

Insemination of the mare

The mare's tail should be bandaged and the anus and vulva washed with clean water and cotton wool. Because of their spermicidal properties, disinfectants must not be used.

When the mare is ready, one of the plastic packs containing the semen is removed from the inner canister of the 'Equitainer'. The lid of the 'Equitainer' must be tightly re-sealed. The pack should be gently rotated to mix the contents − the semen must not be warmed. The correct amount of semen − normally between 20 and 40 ml − is poured into a sterile, all-plastic syringe and inseminated directly into the mare's uterus via an insemination catheter which passes through the cervix.

The second pack of semen remains in the sealed 'Equitainer', which should be stored in a cool place.

Record of insemination

The Mare Insemination Record is a detailed form to be completed by the vet as soon as each insemination has been completed. The mare is described fully, with all markings shown on diagrams, to ensure the valid identity of the recipient of each stallion's semen. The date and time of insemination will be recorded, as well as the quantity of semen used.

Between twelve and twenty-four hours after the first insemination, the vet will examine the mare internally to ascertain whether or not she has ovulated. The second insemination will normally be carried out up to eighteen hours after ovulation. If the mare does not ovulate as expected the stud must be notified so that a fresh batch of semen may be collected and dispatched. Any semen remaining from the initial inseminations must be destroyed by the vet.

All subsequent inseminations must be recorded on the Mare Insemination Record, which must be sent back to the stud along with the 'Equitainer' bucket.

Pregnancy diagnosis

Eighteen to twenty-one days after insemination an ultrasound scan may be taken to determine whether or not the mare is pregnant. This scan may also help to detect a twin pregnancy. Most studs require a Pregnancy Diagnosis Certificate stating whether the pregnancy diagnosis is positive or negative upon examination between forty-two and sixty days after insemination. Upon receipt of this certificate the stud will normally issue their Stallion Covering Certificate.

Whenever stud fees are based on a no foal, free return or no foal, no fee arrangement, the mare must be checked again before 1st October, as this is normally when fees are due for payment. If the mare is not in foal then, the whole procedure must start again.

EMBRYO TRANSFER

The technique of embryo transfer is an exciting and relatively new development in horse breeding. The research in Great Britain has been led by Dr. 'Twink' Allen of the Equine Fertility Unit at Newmarket.

The advantages of removing an embryo from the donor mare

and placing in the uterus of a recipient mare to carry to term are:

The valuable donor mare may, for one reason or another, be unable to carry a foal to term. This may be because of age or damage to the reproductive tract.

The donor mare is normally an outstanding competition horse. Using this technique, her foal may be born without her having to take time out of competition.

More than one foal may be bred per season 'out of' one quality mare.

The main disadvantages are that it is expensive, and some breed societies will not register progeny resulting from this technique.

Technique

Both mares are closely observed and regularly examined and, through hormone treatments such as prostaglandin and chorionic gondatrophin their oestrus cycles are synchronized. This synchronization is essential in order that the embryo is transferred from and to very similar environments. For this reason, more than one recipient mare may be prepared — this allows the use of the mare whose cycle most closely simulates that of the donor mare. It also ensures that there is a recipient for the embryo in the event of unsuitable conditions, such as infection, in another mare.

The donor mare is selected for her outstanding competitive qualities, combined with good bloodlines. She must be scrutinized closely in order to be as sure as possible that she will conceive.

The recipient mares will be chosen on their trouble-free breeding records, suitability of size, and age. The ideal age is generally between four and ten years.

The donor mare is either serviced naturally or inseminated artificially. Once the embryo is seven days old it measures approximately half the size of a pinhead. It is at this stage, when it is well formed but not too delicate, that the transfer must take place.

Both mares are restrained calmly in specially designed stocks, to which they are accustomed through previous use for internal examinations. The donor mare's uterus is filled with a warm, nutrient, electrolyte solution, which flushes out the embryo. The flushing medium is collected in warm, sterile flasks and carefully passed through a very fine filter in order to find the embryo. Once found, the embryo is washed in a special solution to remove debris and then examined under a microscope to check for damage or unsuitability.

The most common practice with horses is to transfer the embryo into the recipient mare immediately, although it *can* be frozen for use at a later date. There are two methods of effecting transfer:

By transfer catheter. The vulva of the recipient mare is washed carefully and rinsed with saline solution. The tail is covered with a protective, sterile sleeve. It is very important that the whole environment is scrupulously clean as any contamination will kill the embryo, and contact with any substance, such as blood, soap, antiseptic or water will cause contamination. The special transfer catheter is inserted, via the cervix, into one of the uterine horns. The plunger is then depressed, thus placing the embryo into the correct position.

Surgically. Under sedation and local anaesthesia, a small incision is made in the flank, the uterine horn is exteriorized and the embryo placed directly into the uterus. The incision is then closed. With this method, there is much less chance of introducing infection, and better results are obtained.

After transfer, hormone injections are given to the mare to help establish the pregnancy. At experienced embryo transfer stations, pregnancy rates of around 65 per cent are obtained.

Fourteen days later, when the embryo is twenty-one days old, a scan should show the progress of the pregnancy. The donor mare should also be scanned to ensure that there is not one of a pair of twins still in the uterus.

Once born, the foal will show the temperament and characteristics of the true genetic parents — and will, hopefully, justify all the time and expense by proving to be a potential superstar!

7

DIAGNOSIS OF PREGNANCY

Attempting to judge whether a mare is pregnant or not through the absence of oestrus is not entirely satisfactory as there are other factors which could contribute to her not coming into season. Some mares return to oestrus but do not show any outward signs. This is known as a silent heat. The mare may not be pregnant, but fails to return in season because of prolonged dioestrus. (Confusingly, some mares show oestrus when pregnant, and may unwittingly be covered. This does not normally harm the embryo unless the cervix is opened during coitus — which is more likely to occur in old or recently foaled mares.)

METHODS OF DIAGNOSIS

Pregnancy can be more accurately diagnosed through the following methods:

Rectal palpation.

Ultrasound scanning.

Blood test and urine analysis.

Rectal palpation

This is a very reliable and cost-effective method of pregnancy detection:

After days 18−21: if pregnant, the walls and horns of the uterus will feel turgid and a foetal sac may or may not be felt in the uterus.

After days 30−35: the foetal sac is enlarging and is therefore easier to feel. Non-adjacent twins can be detected at this stage.

After day 60: if twins have gone undetected they will now be very hard to distinguish. The pregnancy should be checked at this stage, as some are known to fail between days 40 and 60.

Ultrasound scanning (echography)

Ultrasound scanning allows a positive diagnosis to be made as early as twelve days after ovulation, and the identification of twins at this early stage facilitates prompt termination, allowing the mare to be re-covered at her next oestrus.

A probe, housing an emitter and transducer, is inserted into the mare's rectum. Soundwaves of an ultra-short wavelength are emitted and, as these waves travel in straight lines, they are either absorbed, reflected or become much weakened by distance. An electronic process converts the reflected echoes into a visual form − the results of this can be seen immediately on screen.

Blood tests and urine analysis

Hormones present in the mare's blood and/or urine can be detected at different stages of the pregnancy. These hormones include:

Pregnant mare serum gonadotrophin (PMSG) − also known as equine chorionic gonadotrophin (ECG). This is produced by the endometrial cups in the placenta between days 40−120 and can only be detected through blood testing as it is not present in the urine. It is an accurate indication of pregnancy − the sampling 'window' is days 45−95 although, in the case of foetal death occurring after the formation of the endometrial cups, PMSG will continue to be produced − giving a false positive result.

Progesterone. This hormone is responsible for maintaining pregnancy and can be detected in the blood and, in a lactating mare, in the milk as well. If the milk is to be tested, days 18 and 19 are optimal. A positive blood test is only an indication. False results can be obtained as a result of prolonged dioestrus, early foetal death or a very short oestrus cycle. Knowledge of the mare's normal cycle is therefore necessary − although it must be noted that this is not a reliable pregnancy test.

Oestrogen. High levels of oestrogens are produced by the foetal gonads. Levels rise rapidly after 100 days, peak at about days 120−150 and remain high until after 300 days. As the hormone is produced by the live foetus, using the detection of oestrogen in the plasma or urine of the mare is an accurate but late method of pregnancy detection. Blood tests are reliable after 120 days: urine tests are reliable after 150 days.

A negative blood test certificate will usually only be accepted by studs if accompanied by a negative manual examination or ultrasound scan.

TWIN PREGNANCIES

During the routine scanning when testing for pregnancy, a twin pregnancy, if present, will almost certainly be detected. This is undesirable because it almost inevitably results in either the abortion of both foetuses or the ultimate birth of dead or undernourished foals, because the mare's placenta cannot provide adequate nourishment for two foetuses.

Nature plays a role in the control of twins: one embryo may die in early pregnancy, allowing the normal development of the remaining conceptus. This is termed natural reduction. However, the risk of complete pregnancy failure is too great to rely upon this method and, as the abortion of twins can lead to further complications and affect future fertility, the vet will usually try to prevent a twin pregnancy from progressing in the very early stages.

In the event of a twin pregnancy, one of the following may occur:

1) One embryo may die in early pregnancy and be resorbed, allowing the remaining one to develop normally.

2) One foetus may die in a later stage of pregnancy. The live foal may be born undernourished, accompanied by the dead foetus.

3) One foetus may die in the middle/late stage of pregnancy, causing the abortion of both.

4) If both live to beyond day 60 they will compete for space, which may lead to one being trapped at the tip of one uterine horn. This foetus will then die, and may be evident among the membranes after the birth of the other foal.

5) Rarely, both foals go to term and are born alive, though very undernourished. Provided they both have the essential colostrum in the first twenty-four hours and are kept under close veterinary care they should both have a chance of survival. Bottled colostrum may be needed to ensure adequate supplies for both foals. This must be organized well in advance.

Diagnosis of twin pregnancy

One-third of twin pregnancies form in the same uterine horn— this can make detection by manual palpation very difficult, as the swelling feels no larger than a normal single pregnancy of the same age. During the first 60 days, it is easier to detect twins forming in separate horns. After this time the swellings are much closer together − almost united − making it much more difficult to determine the presence of two foetuses.

Detection by ultrasound is much more accurate than by normal palpation, but it is not absolutely reliable. Usually, however, a twin pregnancy will be detected by this method as early as days 14−16 . As with manual palpation, it is very difficult to define two foetuses after sixty days.

Ideally, a twin pregnancy should be detected before day 33 so that action can be taken to terminate either the whole pregnancy or one conceptus. If the whole pregnancy is terminated the mare may then stand a chance of a second pregnancy within that breeding season.

Procedure upon diagnosis

Having diagnosed a twin pregnancy the vet then has to decide whether to intervene and, if so, when. The methods of intervention are:

Manual Rupture. One conceptus, normally the smaller, is crushed. From day 21 the success of this method is reduced. Also, if carried out too late, it can cause uterine trauma leading to the release of natural prostaglandin and, subsequently, complete pregnancy failure.

Prostaglandin treatment. Upon conception, the corpus luteum produces progesterone, which stimulates preparation of the uterus for pregnancy. The hormone, prostaglandin, instructs the corpus luteum to cease its activities. Upon administration of prostaglandin the corpus luteum will be resorbed, causing a decrease in progesterone and making the uterus inhospitable to the conceptuses. They may then be resorbed and the cycle will begin again. This treatment must be given before day 36 because, once the endometrial cups start to produce the hormone ECG, it is difficult to get the mare back into season using progesterone. The subsequent heat will be fairly normal, with average fertility.

Ultrasound-guided injection. One twin is terminated by ultrasound-guided lethal injection into the foetal heart in mid-pregnancy. There is a success rate of around 95 per cent provided that the two conceptuses are separated (either in separate horns, or manually separated). The earlier this procedure is carried out, the greater the success rate.

Whichever course of action is chosen, the most important factor will be the degree of monitoring of the pregnancy through repeated examination by an experienced vet. Obviously this will increase costs incurred.

8

THE STAGES OF PREGNANCY

Pregnancy is a complex process spreading, in the horse, over a period of approximately eleven months. The overall process can be considered as a series of interconnected but distinct stages.

FERTILIZATION

Within four to six hours after coitus the sperm should arrive in the fallopian tubes, where they remain, fully fertile, for at least three days. Ideally the mare will have been covered approximately twenty-four hours before ovulation, so that an ovum on its journey to the uterus will have an increased chance of uniting with a sperm within the fallopian tube.

Sperm attach to the ovum and, through the whipping action of their tails, try to penetrate the ovum's outer layer. The nucleus of the successful sperm unites with the nucleus of the ovum, combining their genetic material and thus determining the characteristics of the potential foal.

Once fertilized, the ovum becomes resistant to the entry of other sperm and is known as a zygote. The zygote descends the fallopian tubes and begins to divide, (the process of mitosis), resulting in a mass of cells which is then referred to as the conceptus (or morula). The conceptus is helped through the fallopian tubes by muscular actions and the brushing action of the delicate, hairlike cilia, which line the tubes. Secretions from the tube provide nourishment for the conceptus.

Non-fertilized eggs remain in the fallopian tubes and degenerate.

DEVELOPMENT OF THE FOETUS AND FOETAL MEMBRANES

On day 5, the conceptus reaches the uterus and, until day 16, floats about freely, nourished by the secretions of the mucous membrane (the endometrium), which lines the uterus.

The cells of the conceptus divide, forming a group of membranes which appear like a fluid-filled ball, within which can be seen a thickened area called the inner cell mass. At this stage, there are three distinct areas:

1) The inner cell mass — which is to become the embryo.

2) The blastocoel (fluid-filled centre), which is to become the yolk sac.

3) The trophoblast, which is responsible for the formation of the membranes which will attach the conceptus to the endometrium.

From day 9, two layers of cells become differentiated: the first wave of cells proliferate and cover the inside of the ball, forming the endoderm. The second wave of cells cover the outside of the ball, forming the ectoderm. These layers of cells spread around the embryo, forming the yolk sac wall and trapping it within a complete bubble of fluid in which it will develop. At day 14 a third cell layer — the mesoderm — begins to develop between the endoderm and ectoderm. The mesoderm spreads around the embryo, giving a distinct division between the yolk sac and the smaller area behind the embryo, known as the allantoic cavity or allantois.

The yolk sac provides nutrition through the trophoblast — it is filled with 'uterine milk' secreted by the crypts of the uterus.

At this stage the membranes are known as extra-embryonic membranes, because they are outside the embryo. All placental membranes originate from them: the ectoderm forms the attachment with the wall of the uterus; the mesoderm forms the blood vessels and nutrient transport system; the endoderm forms the inner cell lining which becomes the allantoic sac.

The allantois develops as a fluid-filled cavity behind the embryo. It continues to expand and virtually encircles the embryo, which is floating within the amniotic cavity, formed by a membranous sac, the amnion. The amnion is formed as a result of the fusion of the mesoderm and ectoderm. The amnionic cavity is the container for the amniotic fluid, in which the embryo will develop.

A small part of the endoderm becomes closely associated with the embryo, narrows and forms a tube which runs the length of the inside of the developing embryo. This tube is called the primitive gut tube. Some other organs develop from this tube.

Blood vessels enter and leave the embryo via the umbilical cord, which consists of special supportive, connective tissue, attaching the embryo to the membranes through which nourishment and waste matter are transported. As the embryo develops so too does a canal leading from the bladder to the allantoic cavity, for the purpose of removing urine. This canal is the urachus.

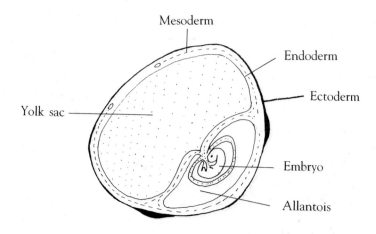

Figure 15 The foetal membranes at approximately day 30

At approximately day 35, the development of all organs is virtually complete and the embryo is now known as a foetus.

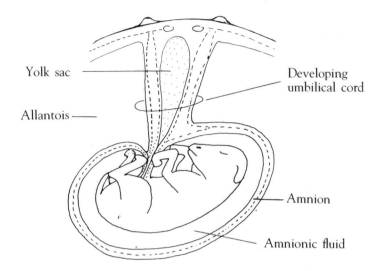

Yolk sac

Allantois

Developing umbilical cord

Amnion

Amnionic fluid

Figure 16 The foetal membranes at approximately day 55

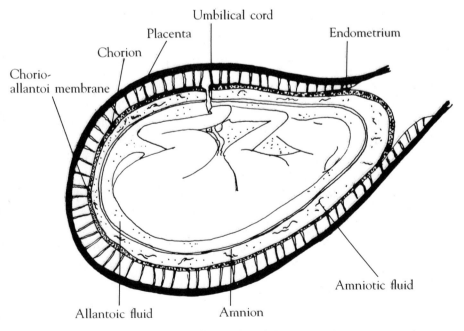

Umbilical cord

Placenta

Chorion

Endometrium

Chorio-
allantoi membrane

Allantoic fluid

Amnion

Amniotic fluid

Figure 17 Well developed foetus in the uterus

Around day 40, cells from the girdle surrounding the foetus migrate to the uterine wall at the junction of the body and horn of the uterus on the side where the foetus is developing. These cells form the endometrial cups, several of which are positioned in a band at the base of the uterine horn. Between days 40 and 100, the endometrial cups secrete the hormone pregnant mare serum gonadotrophin (PMSG), also known as equine chorionic gonadotrophin (ECG). The mare mounts an immune response against the invading cup cells and, by day 90, they begin to regress, disappearing by day 120.

While the embryo is developing into the foetus, in other parts of the conceptus, further contact between the mesoderm and ectoderm results in the formation of another membrane, the chorion. The outer surface of this membrane produces minute, hairlike microvilli which are richly supplied with capillaries — giving the whole surface a red colouring. These microvilli are arranged in millions of small groups known as microcotyledons. On the maternal side (the uterine wall) there are also many millions of microcotyledons in the form of microscopic capsules. A tiny arterial blood vessel enters the top of each capsule before forming a capillary network. This, in turn, becomes a venous blood vessel, which takes blood away to the lower end of the capsule.

The foetal microcotyledons 'button' into the maternal ones, forming a very close association which facilitates the exchange of nutrients and waste products through gaseous exchange and diffusion. Upon birth, the microcotyledons 'unbutton', allowing complete expulsion of the membranes.

In the meantime, the allantois forms a complete sac, the allantoic cavity, around the foetus. All foetal waste matter is excreted into the allantoic cavity where some of it solidifies, forming the brown, jelly-like hippomanes, which may later be seen amongst the afterbirth.

Within the allantoic cavity is the allantoic fluid or 'the waters'. This fluid is composed of water, urea, albumin, lactic acid and various salts. The allantoic fluid serves several functions:

It offers protection to the foetus from knocks and bumps.

It allows a degree of movement within the uterus.

It cushions the uterus against the vigorous movements of the foetus.

At birth, the escaped fluid helps to lubricate the birth canal.

The inner lining of the allantois is permeated by many blood vessels. It is very shiny, and it is this shiny surface that is seen when the membranes are expelled as the placenta is pulled inside-out during the third stage of labour (the afterbirth).

The chorion and allantois are so closely associated that they are often referred to jointly as the chorio-allantoic membrane (CAM), the allantochorion, or the chorioallantois.

The allantoic fluid separates the CAM from the third membrane, the amnion. This glistening white membrane, which completely encapsulates the foetus, contains the amniotic fluid, an alkaline solution of water, proteins, urea, sugar, lactic acid, salts and keratin, of which there is eventually about 5−6 litres (9−10½ pints). The foal is born within the amnion.

MILK PRODUCTION

Milk is an essential substance. In its first form (colostrum), it provides the newborn foal with both nutrients and protective antibodies. Subsequently, it provides the staple food for the young foal. The developments which lead to milk production begin during the final month of pregnancy.

The mammary glands

A mare has four mammary glands, in two pairs. Each pair is situated in one half of the udder, separated and supported by a dense, fibrous wall − the medial suspensory ligament − which runs along the mare's midline. The glands are given extra support by lateral suspensory ligaments running under the skin, and by strands of supportive tissue which extend from this ligament into the mammary tissue.

The mammary tissue consists of millions of alveoli held in bunch-like arrangements (lobules) by connective ducts. Groups of lobules are linked to form lobes, from which ducts extend and lead to the gland cistern. Each gland has its own gland cistern and teat cistern. The secretions of each pair of glands exit via a single teat. The extremity of the teat is closed to prevent leakage — there is no organized sphincter muscle.

During pregnancy, increased levels of progesterone cause increased development of mammary tissue, and an increase in prolactin, growth hormone and cortisol stimulates milk production (lactogenesis). Milk production starts in the last two to four weeks of pregnancy, the milk being synthesised in the lactating cells which line the alveoli.

The composition of milk

The first milk produced is the nutrient-rich colostrum, which is produced for approximately three to four days. Colostrum is a natural laxative — it stimulates peristalsis (muscular contractions of the gut wall) by promoting secretions from the intestinal glands. Peristalsis helps to clear away the accumulated faecal matter (meconium) and make way for the newly digested food (milk).

The most important function of colostrum, however, is to provide the foal with antibodies, essential as protection against infection. Colostrum contains at least 13.5 per cent protein, in the form of various protein immunoglobulins and their associated antibodies. It also contains lactose, calcium and a low level of lipids.

Once the colostrum is finished, normal milk is produced. While this contains a lower protein concentration and comparatively low lipid levels, it also contains a high level of lactose, together with sodium, potassium, calcium, magnesium, phosphorous, iron and water.

Once the secretion of milk is established, it is maintained as a response to the stimulation of the foal sucking. The sucking action provides a nervous stimulus, resulting in the release of the hormone oxytocin from the posterior pituitary gland. Oxytocin promotes milk secretions. The average mare produces around 10 litres (2 gallons) of milk per day.

THE STAGES OF PREGNANCY — A BRIEF SUMMARY

Times and durations given are approximate.

Day 5
The conceptus reaches the uterus. It remains mobile within the uterine horns and body until day 16.

Day 16
The conceptus becomes lodged in one horn of the uterus.

Day 21
The conceptus now measures approximately 1.5—3 cm in diameter. The mass of cells divide and differentiate into those which are to form embryo and those which are to form membranes. Development at this point is rapid, with the cells becoming organized into the various organs (organogenesis).

Day 35
Most organogenesis now complete — the embryo is now a foetus.

Day 60
Foetus and associated membranes are now approximately 12 cm in diameter and fill one horn of the uterus.

Days 70—80
Foetus begins to fill the body of the uterus — the hindquarters remain in the horn.

Day 90
The whole uterus is now filled with fluid, with the distinction between the body and horns becoming less obvious.

From
4—6 months
Foetus developing within the body of the uterus.

From
6—7 months
The foetus is now accommodated by the body and horns of the uterus as it is too large for the body only. After this stage, the foal cannot move to alter presentation.

From
8—11 months
This is the period during which rapid developments occurs — physical movement may be noticed. At some stage during the final month, milk production begins. The length of a normal pregnancy is 330—345 days, although there are often slight variations.

9

CARE OF THE IN-FOAL MARE

This important topic can be considered under two main headings, nutrition and general health care, the overall aims being to maintain the physical and mental well-being of the mare, and to produce a healthy foal.

NUTRITION

The nutrition of the in-foal mare is a very important issue, which has three main objectives:

1) The birth of a healthy, well formed foal.

2) Satisfactory post-natal growth of the foal through a plentiful supply of milk.

3) The maintenance of the mare's correct bodyweight and condition.

At the beginning of pregnancy, the mare should be physically fit and, while looking sleek and rounded, must not be overweight. This physical condition must be maintained throughout the pregnancy. Excessive fat may lead to difficulties in foaling and associated circulatory problems, such as laminitis.

The feeding of each individual mare will depend upon the following factors:

Date of conception. An early conception date in February should result in a January foal. The mare will not enjoy the benefits of spring grass and, because of the inclement weather, may spend a lot of time stabled. A later conception date in June should result in a May foal. The mare will be able to spend longer periods grazing, enjoying better weather and the spring grass — normally!

Whether there is a foal at foot. If the mare was covered with a foal at foot she will be pregnant and lactating simultaneously, which will put extra demands on her.

Breed. When breeding Thoroughbreds early foals are desirable, as they are registered on 1st January and consequently race against youngsters of the same age, (horses born in the same year). Therefore Thoroughbred mares do not benefit from the grass in the latter stages of pregnancy and may spend long periods stabled if the weather is very bad.

Non-Thoroughbred types tend to foal later and will be fed concentrates depending on the quality of the grass available. Native ponies, however, often need no extra concentrates as they 'do well' and have a tendency to put weight on easily.

Feeding in the first seven months

During the first two-thirds of the pregnancy, the foetus does not grow so rapidly as later, so the mare will simply need to be fed a normal, balanced diet of good quality hay and concentrates. As mentioned earlier, the main concern is that the mare should not become over- or under weight. Therefore, actual quantities will be dependent upon her type, temperament, the time of year and work (if any) being done. The mare with a foal at foot will not usually be working.

The growth of the foetus will, however, be affected by the quantity and quality of whatever the mare eats, and great harm — possibly abortion — may result from the feeding of mouldy food because of the toxins produced by fungi.

As with all horses, the mare's diet should contain the correct balance of nutrients, with particular attention paid to the provision of the following:

Protein. This may be provided in grass, grass meal, soya bean meal, milk powder, protein nuts and stud cubes.

Calcium. From limestone flour supplement.

Phosphorous. From cereals.

Vitamins A and D. From green foods, cod liver oil, sunlight.

Vitamin E. From fresh foods and cereals.

Folic acid, vitamin B12, manganese and zinc. Mainly from vitamin and mineral supplements.

Water. A constant supply of clean, fresh water must always be available.

If there is any doubt as to whether the mare was receiving the correct balance of nutrients, a metabolic profile may be taken from the analysis of a blood sample. Any deficiencies may then be corrected in the early stages of pregnancy.

Calcium and phosphorous requirements are 20 g and 14 g respectively for a 500 kg mare during the first third of pregnancy; this will be supplied by good quality forage. These requirements increase by 85 per cent, to 37 g (Ca) and 26 g (P) in the last third of pregnancy. These levels cannot be supplied by increased feed alone, and require supplementation. More generally a good mineral/trace element supplement should be fed to pregnant mares, especially during the first and last thirds of pregnancy, as these are times when foetal stores of trace elements are being laid down.

Feeding in the eighth to eleventh months

Energy and protein requirements increase by 20 per cent and 32 per cent respectively above maintenance levels during the last third of pregnancy. For a 500 kg mare these increases can be met by 1 kg of good quality grain mixture fed daily. However, sudden changes in diet should always be avoided, especially in the latter stages of pregnancy.

The daily ration of the pregnant mare should contain at least 12 per cent crude protein. The rations may be calculated by referring to the relevant section in another book in this series *The Horse: General Management*. The protein in the diet should be of a high biological value, that is to say it should contain a high level of essential amino acids.

Stud cubes are specially designed to provide the mare with a balanced diet, with the correct levels of protein, vitamins and minerals. The essential amino acids, lysine and methionine, are added to these cubes to ensure optimum growth and development of the foetus. (Stud cubes are also used to feed stallions, as the balance of vitamins helps to maintain fertility.) As with all feedstuffs, the stud cubes must be introduced to the mare gradually, so that she is on the full ration when the final three months of pregnancy commence.

As the foetus develops it occupies more room in the abdominal cavity, making less room for bulky feeds. Therefore feeding large quantities of hay or oat straw should be avoided, since it may lead to an impaction within the large intestines. The hay that is fed should be of good quality, and thus easily digested. Coarse, stemmy hay will take much longer to break down. If no high quality hay is available, feed an alternative such as dried alfalfa.

Try also to feed succulents, which have a slightly laxative effect and help to keep the intestinal contents 'on the move'.

GENERAL CARE

Throughout pregnancy, records must be kept in respect of all general health checks. In addition to monitoring the mare's general welfare, much of the information will be required by the stud. Matters for consideration will include:

1) Vaccination records. The mare should be vaccinated against influenza and tetanus one month before foaling. Many studs wish to see the up-to-date vaccination certificate. It may also be necessary to vaccinate against the equid herpes virus, rhinopneumonitis.

2) Parasite control. The aim should be to reduce the mare's current parasite burden and prevent contamination of the paddocks with parasite eggs. Ivermectin and Pyrantel are probably the wormers presently of choice for parasite control. Slow rotation (continual use for six months or more) is recommended to minimize resistance problems. Mares should be wormed every four to six weeks, even in late pregnancy. Strongyloides westeri (threadworm) infection of the newborn foal via the mare's milk can be reduced by worming the mare on the day of foaling.

3) Feeding. The stud will need to know what the mare has been eating at home to ensure a similar diet or a smooth transition from one diet to another.

4) Behaviour. How does the mare behave when out with other mares? Are there particular problems with covering?

5) Foaling history. Is this a first foal, or have there been difficulties with any previous foalings?

6) Dates and results of swabbing (required for certification).

7) Date of covering. It is quite normal for mares to be covered at least twice.

When the mare goes to stud to be covered she should arrive at least a week before she is due in season so that she may become accustomed to the different environment and routine. If she is foaling at stud and is then to be covered during the foal heat, she should ideally arrive at stud about one month prior to foaling. This enables her to build up appropriate antibodies to the local conditions to pass on to the foal in the colostrum.

Before arriving at the stud, the mare's feet should be well trimmed and she must be unshod behind. Also, teeth should be checked, and rasped if necessary.

The early months of pregnancy

It is during the early stage of pregnancy that there is a risk of the embryo being resorbed. However, although the mare should be kept as free of stress as possible and never be at risk of falling, she is quite capable of doing light work. If the mare is out at grass she will be able to exercise herself but, if stabled, she will need to be exercised ridden, in hand or on a horse walker. The object is to keep her physically fit, particularly in the heart and abdominal muscles — and lack of exercise may lead to fluid-filled swellings of the hind limbs.

The middle months of pregnancy

The mare is able to remain in moderate work until the third or fourth month. Increasing attention must be paid to the diet, ensuring that it is well balanced and keeping the mare in good condition without being fat. Higher levels of protein should be introduced gradually into the diet.

The final months of pregnancy

During the final three months of pregnancy the mare, although not being subjected to undue stress, must still have daily light exercise and must be on a full 12 per cent protein diet. The diet should be an easily digestible one, with very slight laxative qualities to prevent impaction. As previously discussed, a balanced diet is essential: under-nutrition in the last month of pregnancy may affect the colostrum and milk yield, as it is during this time that the udder develops.

The mare should be allowed to become accustomed to the foaling box.

PROBLEMS IN LATE PREGNANCY

'Running milk'

Some mares habitually 'run milk' during the last two to three weeks of gestation. This *may* be a sign of placentitis (see below), premature delivery, or abortion, but it may also occur in mares who are otherwise quite normal. 'Running milk' is a cause of substantial loss of colostrum and its essential antibodies.

Therefore, mares who lose more than a small amount of milk should be 'stripped out' and the colostrum stored by freezing. The colostrum can then be carefully thawed for administration to the foal within the first few hours of life.

Placentitis

Bacteria and fungi can cause placental infection. This is often predisposed by poor vulval conformation and pneumovagina (vaginal windsucking). Signs, which include vaginal discharge, premature mammary development and lactation, usually occur in the eighth to tenth months of pregnancy.

Approximately 10 per cent of abortions are caused by placentitis. Alternatively, foals may be born weak, infected, premature, or be stillborn. Therefore, any discharges will have to be examined to identify the microbe responsible before appropriate treatment can be given. In the case of a mare with faulty vulval conformation, vulval suturing (Caslick's operation) will have to be performed to help reduce the incidence of placentitis.

Colic

Foetal kicking in late pregnancy is believed to cause signs of mild colic. However, colic may also be a sign of pre-term parturition or abortion, or some other more serious conditions. Therefore, any mare with more than very mild signs of colic, which pass quickly, should be seen by the vet.

Prolonged pregnancy

Approximately 1 per cent of pregnancies continue for longer than 370 days. So long as there are no signs of ill health, prolonged gestation is not an indication that parturition needs to be induced.

10

FOALING

The actual process of foaling takes place in three distinct stages of labour. Although most foalings are relatively straightforward, any complications can be avoided or minimized by sound preparation.

PREPARATION

Foaling at grass

Foaling is a natural process, and foaling at grass is obviously what nature intended. However, it is not always practical — particularly if the weather is poor. Other drawbacks are:

Observation is difficult, especially at night.

There is always a danger that the mare may foal near a fence — a particular danger if the fence is wire or if there is a ditch on the other side.

If the field is not entirely free from hazards, foals risk entrapment, injury, or death (for example, becoming trapped under a water trough, or drowning in a pond).

If there are other horses in the field, they may become highly curious, and perhaps knock the foal about, causing great distress to both foal and mare.

If, therefore, through either choice or necessity, the foaling is to take place at grass, it should do so in a safe, empty field. Since most foalings take place at night, the provision of some form of lighting is highly desirable.

The foaling box

If foaling is to take place under cover, a suitable box must be prepared. This needs to be very large to ensure that, whichever way the mare lies stretched out, there is sufficient room behind her. 5 m × 5 m (16 ft × 16 ft) is a suitable size.

The box must be thoroughly disinfected and well bedded down — a deep layer of straw helps prevent slipping and, unlike shavings, will not stick to the foal.

All additional fittings such as mangers and bucket holders must be removed. Although fresh drinking water should be available, it is best provided in a strong bucket without a handle. Hay must be fed on the ground, not from a net or rack.

Ideally, a dimmer switch should be incorporated in the light fitting. An infra-red lamp may be useful after the foal has been born.

A facility for monitoring the mare as unobtrusively as possible is essential. In large studs, a closed-circuit television is often used. Additionally or alternatively, a room adjacent to the foaling box with an observation window is very useful. A foaling alarm, placed around the mare's neck, will set off an audible warning in the observation box as soon as she begins to get hot. Alternatively, an ordinary radio-linked baby alarm may prove useful.

Other preparatory procedures

If foaling at home, as first stage labour (see next section) commences alert the vet to be on standby. Hot water, soap, nailbrush and clean dry towels must be available for handwashing. Also have ready:

Sterile foaling ropes in a sealed pack. (As a precaution: the vet will normally bring ropes for use in awkward foalings. Also note that unqualified persons should not attempt to pull out a badly presented foal.)

Towels to rub down foal if necessary (if the mare is too tired or weak to lick, or if it is extremely cold.)

Obstetric lubricant.

Bucket, in which to put membranes.

Iodophore liquid for the umbilical cord.

Baby's feeding bottle and teat.

Frozen colostrum and powdered mare's milk replacer.

String.

Disposable rubber gloves (sterile).

Heat lamp.

The mare's tail should be bandaged to aid hygiene and keep it clear of fluids. If the vulva has been stitched to prevent vaginal windsucking, the stitches must be removed by the vet before the onset of second stage labour. On occasions when these stitches are no longer present, it may be preferable for the vet to enlarge the aperture rather than risk excessive tearing. This procedure is known as episiotomy. Since the vet must be allowed sufficient time to carry out such procedures, they should be considered preparatory.

It must be stressed that handlers should interfere as little as possible throughout the foaling process, but should watch quietly from outside the foaling box.

It is also important to appreciate that second stage labour is a rapid process in the mare — if complications arise, the chances of the foal surviving are low. It is therefore important that someone experienced is on hand, who can recognize quickly when things are going wrong. Thus, for an inexperienced breeder, it is safer to send the mare to stud to foal.

In order to gain experience, try to visit a busy stud farm and watch plenty of foalings — knowing what is normal helps in the recognition of abnormal occurrences.

LABOUR

First stage labour

This is mainly a preparatory stage, which usually lasts between one and four hours. However, there are no hard and fast rules as to what to expect. Signs may include one or more of the following:

Slackening of the sacrosciatic ligaments, causing the hind-quarters to appear weak. This slackening of the ligaments facilitates widening of the pelvic girdle to allow an easier passage during birth.

Relaxation of the vulva.

The udder fills and may leak milk.

'Waxing up' as a coating of honey-like colostrum covers the ends of the teats.

Restless behaviour: looking around at flanks and shifting weight from one hind leg to another; digging up the bedding — mild, colic-like signs.

Increase in temperature — sweating.

As the labour begins to progress, the mare may grunt and start to lie down and get up repeatedly.

Every mare differs in the signs shown and when they are shown — there is always the chance of false alarms as these signs may become evident days before the actual foaling. The waxing up is generally taken as a good sign that the mare is in first stage labour.

During first stage labour, the muscles of the uterus start to contract in waves. The direction of the contractions, from uterine horn to cervix, help to position the foal in preparation for birth.

Throughout late pregnancy the foal lays ventrally (upside down), with the spine towards the mare's abdomen and the forelimbs flexed. During first stage labour the body rotates into a dorsal position (spine uppermost) with the head, neck and forelimbs extending forwards toward the vulva in a 'diving' position. The cervix gradually dilates in response to uterine contractions and pressure from the foal's forelimbs. The vulva relaxes further and vaginal secretions increase.

Sometimes, there is ventral anterior presentation. This means that the foal is pointing the correct way, but is lying back downward. Occasionally a safe delivery is possible without manipulation.

Second stage labour

This stage covers the expulsion of the foal. It normally occurs at night and, in a straightforward delivery, takes approximately twenty minutes — although it can range from five minutes to an hour.

At this point, the mare is normally lying down. The cervix opens, allowing the CAM to bulge into the vagina. Pressure causes the CAM to rupture, releasing allantoic fluid. This heralds the start of second stage labour. Early on in this stage the position of the foal should be checked. After this point, matters should proceed in distinct stages:

1) The mare will strain, using her outstretched limbs and abdominal muscles.

2) The glistening white amnion should become visible within about five minutes of the CAM rupturing. This may contain fluid and one foetal foot.

3) Both forefeet may be seen — one is normally further forward than the other. The nose should be just behind the feet. An attendant may check quietly to ensure that all is normal at this stage. If there are any abnormalities — for example, no sign of the head — the vet must be called immediately. An *experienced* stud hand can assist while waiting for the vet. Normally the foal's feet will rupture the amnion, allowing the passage of air. If this does not happen the attendant may have to break the amnion and clear the nostrils of mucus so that the foal is able to breathe.

4) Once the head has emerged the mare may get up, rest or roll. Nervous mares may get up if disturbed.

5) The mare will strain again and the chest and hips will appear. She may then lie for a while with the hind limbs in her vagina. During this time there is a vital transfer of the mare's blood to the foal through the umbilical cord. The foal should start to breathe as a result of a first-ever contact with cold air. If the foal does not start to breathe immediately action must be taken (see Problem Foalings).

6) The hind limbs will shortly come free of the vagina and, as the foal moves, the amnion will rupture completely.

7) Further movement, as a result of the mare getting up or the foal struggling, causes rupture of the umbilical cord close to the foal's abdomen. *Never cut the umbilical cord* — this causes complications and the risk of infection. Once the cord has ruptured, it should be dipped in dilute iodine solution.

8) At this point the vital bonding occurs as the mare licks the foal to dry the coat and stimulate circulation — leave them alone.

At an early point after the initial bonding, the foal's navel stump may be dipped in iodophore to prevent infection. (Never dip the navel stump in anything strong, such as undiluted iodine or disinfectant.) The mare will normally have the amnion and umbilical cord hanging from her vulva. These should be tied, to keep them off of the ground. They should never be cut, as they provide the natural traction needed to cause separation of the CAM from the endometrium. Never apply enforced traction to the amnion and cord, as this may result in a rupture of the membranes or prolapse of the uterus.

Third stage labour

This involves the expulsion of the membranes and is also known as cleansing. It normally occurs within three hours of birth. If it has not occurred within ten hours the vet must be

called, as serious problems such as laminitis or metr
result.

In third stage labour, the mare may show signs of ab
pain, caused by renewed uterine contractions. The weight of
the amnion, pulling on the CAM via the umbilical cord,
causes separation from the endometrium. The CAM is turned
inside out as it is expelled. The membranes must be kept and
inspected to ensure that they are complete. Place them in a
water bucket for possible future inspection by the vet in the
event of any complications.

Inspection of the membranes

Separation of the CAM and endometrium normally occurs
rapidly. When laid out, they should reflect the shape of the
uterus — that is, have a body and two horns. Occasionally, the
placental horns may be ripped, resulting in a torn section of
membrane remaining inside the uterus. The consequences are
the same as for retained placenta, hence the importance of
checking that the membranes are complete. The most common
portion to be retained is the tip of the non-pregnant horn.

PROBLEM FOALINGS (DYSTOCIA)

Most foalings are straightforward — dystocia, which means 'bad
birth', is the term used to describe any problem which interferes
with the normal birth of the foal. Dystocia is always a serious
problem, as it may lead to the death of the foal and, possibly,
the mare as a result of one or more of the following:

1) Pressure or twisting of the umbilical cord will cut off the
   supply of blood and, therefore, oxygen to the unborn foal.
   (Especially relevant in posterior presentations).

2) During labour, the placenta separates rapidly from the endometrium. This causes loss of the oxygen supply to the
   unborn foal.

3) The mare may continue to strain even though the foal is stuck. This causes damage to her reproductive tract, possibly leading to peritonitis (inflammation of the lining of the abdominal cavity) or haemorrhaging. The effort of straining may also lead to exhaustion so that, even if the foal's position is corrected, expulsion cannot proceed.

4) As a result of excessive straining the uterus may prolapse (become partially or completely displaced and therefore visible through the vulval lips). This is, however, rare in equines.

5) Rectovaginal perforation may be caused by the foal's foot being driven through the rectal or vaginal wall as a result of the mare's expulsive efforts.

Recognition of dystocia

After the first few minutes of second stage labour, there should be signs of the fluid-filled amnion appearing through the vulval opening. Within the amnion there should be at least one foetal foot − the other should be behind it and the nose behind that. If this is not the case the vet or stud attendant must examine the mare to determine as quickly as possible why this is so. As previously mentioned, it is important that an experienced person is on hand to recognize any abnormalities quickly.

The position of the foal can be checked when the mare is lying down or standing. The mare's perineum should be cleansed and the attendant or vet should wash their arm with warm, soapy water and/or put on a sterile, well lubricated, disposable glove before inserting a hand into the vagina. When the foal is presenting normally the muzzle and two forefeet should be felt. If this is so, the mare should be left to deliver unaided, under unobtrusive observation. If it is not the case, then it is likely that a specific complication will have to be addressed.

SPECIFIC COMPLICATIONS

Abnormally thick placenta

As a result of abnormally thick membranes, the placenta may not rupture naturally to release the allantoic fluid which heralds the start of stage two labour. Instead, the mare becomes increasing uncomfortable and begins to strain even though the waters have not broken. This causes the intact placenta to bulge through the lips of the vulva. It can be identified as a red membrane with a star-shaped scar in the centre of the bulge. This, known as the cervical star, is the thinnest area of the placenta and is where rupture normally occurs. The membranes will have to be ruptured manually, or by using scissors, before stage two labour can commence.

Forelimb and/or head and neck malposture

One or both forelimbs may be bent back at the fetlock, knee or shoulder, or the head and neck may be flexed abnormally. These positions prevent the foal from passing smoothly through the birth canal and can normally be corrected through manual manipulation. The foal's life is not endangered, as the umbilical cord is still intact and free from pressure.

Hind limb malposture

Should the hind limbs be presented forwards they may get wedged on the pelvic brim, preventing passage out of the birth canal. This *does* endanger the foal, as the chest is lodged within the birth canal and the cord may be squeezed, cutting off vital supplies of blood and oxygen. Therefore, the hind limbs will have to be pushed back manually and traction may be needed to assist the delivery.

Posterior presentation

This describes a foal who is 'back-to-front' (facing the wrong way). If the hind limbs are outstretched towards the vulva

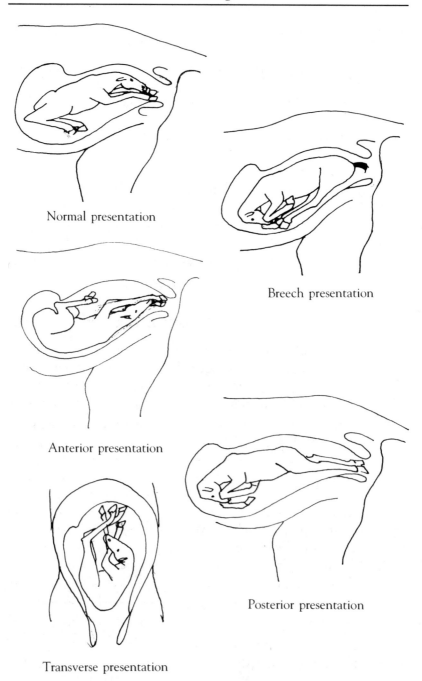

Normal presentation

Breech presentation

Anterior presentation

Posterior presentation

Transverse presentation

Figure 18 Normal and abnormal presentations

there is a chance that, with strong traction, the foal may be born alive.

A 'breech' presentation is a posterior presentation where the hips are flexed and the hind limbs point away from the vulva. This is a very difficult delivery, always requiring veterinary assistance. The first sign may be the presentation of the foal's tail in the vagina — such presentation invariably requires caesarian section.

Transverse presentations

These are very severe malpresentations resulting in the foal lying across the birth canal — the spine is normally very bent, with the feet either pointing away from the vulva (dorsal transverse presentation) or towards the vulva (ventral transverse presentation). As soon as this presentation is diagnosed the mare will normally be transported immediately to the nearest surgery.

Non-surgical treatment

The first step generally taken by the vet is to ascertain the exact position of the foal in order to correct the abnormalities, if possible. The mare is normally unworried by manual examination per vaginum, as she is too concerned with the actual foaling. Because of the pain she may be reluctant to stand still: she may attempt to roll, which could help put the foal into a more viable position.

The vet may tranquillize the mare, which will make manipulation easier. It can, however, make her straining less effective once the foal is in a position to allow delivery.

If all of the mare's natural fluids have gone, warm water with obstetric lubricant may be used to moisten and lubricate the foal. Any parts of the foal which are in the vagina may have to be pushed back into the uterus to allow ropes to be attached as necessary, although simple flexions of joints may be resolved by hand.

Once the foal's limbs have been rearranged into a satisfactory position, traction may be applied to the ropes or limbs as directed by the vet. The mare often assumes a standing position whilst traction is being applied, therefore the foal must be supported while being delivered, to prevent further injury or premature rupture of the umbilical cord.

Surgical treatment

If all non-surgical attempts fail, the vet will prepare for surgery. These presentations are normally only resolvable through caesarian section or embryotomy. If contemplating caesarian section, one of the main considerations will be the effect of the anaesthetic on both the mare and foal. Post-operative complications can include shock, uterine and vaginal haemorrhage, infection and laminitis.

Embryotomy, which involves the removal of a dead foal piece by piece using roughened wire or a knife, may be used when there are no alternatives, for example, when a mal-presented, dead foal is lodged in the uterus/birth canal.

11

CARE OF MARE
AND FOAL

Although birth is an exciting time for the owner, friends and family, the mare and foal should not be disturbed more than is absolutely necessary. During the first hours, the important bonding continues.

CARE IN THE FIRST HOURS

As the foal is born the amnion should rupture naturally. If this has not occurred it must be ruptured manually and a check made that the foal is breathing — the nostrils must be free of mucus. If the foal is not breathing, it will be necessary to administer expired air resuscitation by breathing very positively into the foal's nostrils (mouth to nostril). This is done as follows:

Pull the tongue out, remove all mucus, etc. from the back of the mouth with the fingers.

Try to stimulate a sneeze or cough by tickling the inside of the mouth with a piece of straw.

Hold the head straight, preferably with the body slightly higher than the head.

Blow down the nostrils and fill the lungs, let the foal expire, and repeat.

Check that the heart is beating by feeling the chest wall behind the elbow. If the heartbeat is weak or absent, thump the chest lightly on the sternum several times and positively massage the legs, with rapid circular movements. Oxygen may be administered from a cylinder via a mask if the vet is present.

In the first minute of life the foal's heart rate will be in the region of 60 beats per minute. This increases to approximately 140 beats per minute as the foal struggles to stand and suck. It should then settle to around 75—95 beats per minute and remain at this rate for the first year of life. In a yearling the pulse rate will be around 50 beats per minute.

During the first hour or so, the foal will breathe rapidly, settling to a rate of 20 to 30 inhalations per minute. Body temperature should be approximately 37.5°C (99—100°F).

The mare's natural instinct to lick helps to dry and warm the foal, stimulate circulation and form a bond between herself and her new offspring. If she is reluctant to lick the foal a little salt rubbed in the coat may encourage her. They should then be left in peace while they get to know each other. If, for any reason, the mare is unable to lick the foal, he will need to be rubbed carefully with a clean towel.

At a convenient moment the ruptured umbilical cord may be dipped in iodophore.

Once the membranes have been expelled, the mare's vulva may be carefully cleansed and, if necessary, stitched. These procedures help to prevent the entrance of bacteria, thus reducing the risk of infection.

The foal will generally have a serious attempt at standing within one and a half hours. He will be wobbly at first, but no help should be given — no matter how tempting it may be! However, assistance may be needed if he is still unable to stand after two hours. Try to ensure that there are no projections, buckets etc. onto which the foal may fall.

Suckling

Although the sucking reflex is evident about five minutes after birth, sucking cannot occur until the foal is standing successfully.

After approximately an hour and a half, the foal should be sucking. He will have to search for the teat, but should be allowed to do so unaided if possible. Not all mares are immediately maternal — some may refuse to suckle the foal. Although few mares genuinely reject their foals, nervous mares are made worse by excessive human intervention. In such cases, try turning off the lights and leaving mare and foal alone. Only intervene if the foal is at risk of injury or it becomes apparent that there is a real problem.

It is, of course, essential that the foal sucks in order to take the colostrum. A normal foal will generally suck every half hour or so, drinking approximately 20–30 cl. If the mare refuses to suckle the foal even when restrained, or if the foal is too weak to stand, the mare may have to be milked so that the foal can be bottle fed.

If the mare produces no milk, or dies during parturition, the National Foaling Bank, based in Shropshire, should be consulted. They offer an emergency delivery service of replacement mare's milk as well as keeping a register of any recently orphaned foals, and mares who have lost their newborn offspring. The vet will know who to contact in the case of such emergencies.

Provision of adequate colostral immunity ('IgG')

Good colostrum management is essential if the foal is to have maximum benefit from the antibodies contained within it. The aim should be for the foal to ingest one litre of good quality colostrum, the majority of which should be absorbed within six hours of birth. No more will be absorbed after eighteen to twenty-four hours.

If, in the days prior to foaling, the mare 'runs milk', even at only a 'drip' rate, large quantities of colostrum will be lost. In such cases, rather than losing the colostrum, the udder should be stripped (milked) and the colostrum frozen for administration to the foal after birth.

If, for any reason, there is any doubt as to whether the foal has had sufficient colostrum, (for example if he is very weak, or the mare would not let him suck), the vet can take a blood sample. If the result is low (less than 4 g/l), the foal is at greater risk of succumbing to infections such as joint-ill, pneumonia, etc. This situation will have to be corrected by plasma infusion.

Prevention, (good management, storage of frozen colostrum) is therefore far more desirable and, with forward planning, should be quite possible.

Passing meconium

Meconium is dark brown, black or greenish dung stored in the foal's rectum, colon and caecum while the foal is in the uterus. As previously mentioned, the ingestion of colostrum stimulates peristalsis, which promotes the expulsion of the meconium. Meconium should start being passed in the first twelve hours, and should be completely evacuated within four days. After this, yellow-coloured milk dung will appear.

If this does not happen, the vet or experienced attendant may administer a soap enema to help get things moving. A well lubricated finger inserted carefully into the rectum may dislodge any blockage sufficiently to clear the passage. Signs of meconium retention are that the foal squats and strains repeatedly without passing anything, and may exhibit colic-like symptoms, rolling and sweating.

General care

In the few hours after foaling, the mare may appreciate a warm bran mash. Add a little limestone flour to boost her calcium levels. Fresh-cut grass or soft hay may be given.

Remove any droppings from the stable and ensure that both mare and foal are warm enough — rugs may be needed, depending on the time of year and the weather.

Mare and foal may then be left in peace — although discreet observation will be necessary to ensure that all is well.

The foaling box must be thoroughly scrubbed and disinfected before being used by another mare.

Table 4. Summary of Newborn Foal Adaptation

Time elapsed since delivery	
1 minute	Rapid chest and abdominal movement, preceded by gasps. Breathing rhythm of approximately 70 breaths per minute. Heart rate approximately 60 beats per minute. Temperature 37.5 °C (99−100 °F).
5 minutes	Hind legs withdraw from vagina; lifting of head; extension of forelimbs; blinking − may whinny. Suck reflex present. Takes up sternal position.
15 minutes	(Cord normally ruptured.) Forelimbs stretched out, hind legs flexed under body, first attempts at standing. Breathing rate approximately 50 breaths per minute. Heart rate approximately 140 beats per minute.
30−90 minutes	Foal stands, attempts to locate udder.
1−3 hours	Foal sucks, starts to pass meconium, follows mare.
12 hours	(Membranes to be fully expelled by now.) Breathing rate approximately 30 breaths per minute. Heart rate 80−120 beats per minute. Temperature 38 °C (100−101 °F). Passing meconium and urine. Getting up and lying down. Following mare and sucking at 15−30 minute intervals.
2−4 days	Meconium completely evacuated: milk dung appears.

POST-FOALING PROBLEMS IN THE MARE

Retained placenta

If the membranes have not been completely expelled within ten hours, the vet will have to remove them. The mare must be restrained in view of the foal, who will need to be kept away from the mare's hind legs.

Generally, the vet will insert a well washed arm and try to remove the membranes manually. If easy separation occurs there should not be any haemorrhaging. At no time can any traction be applied to the membranes as they may tear. If manual removal proves too difficult, other treatments will be necessary. Large volume flushing of the uterus may help to

remove toxins and the debris of degenerating membranes. Antibiotics may be given systemically, or into the uterus with
pessaries.

The administration of oxytocin causes further uterine contractions, which may result in the expulsion of the membranes.
After any complication during third stage labour, the uterus
must be cleansed daily with a warm saline solution. This should
help to remove remnants of the membranes.

Gentle exercise (turning out in a small, sheltered, nursery
paddock) helps to expel the fluid. The expelled liquid should
always appear clear — if there is any purulent exudate it would
indicate metritis. In such cases, the vet will decide which
course of action is suitable.

Disorders of the mammary glands

Mastitis

This is a very painful condition in which one or both sides of
the udder becomes inflamed through the effects of either bacterial or mycotic (fungal) pathogens. It is likely to be seen after
weaning, especially if this has been done very early. The
inflammation may involve either the secretory cells of the
mammary glands, or the connective tissue, or both. The quality
and quantity of milk produced will be affected: it may vary
from slightly watery with yellow flecks or clots, to a watery,
blood-tinged secretion.

Signs
The mare may appear generally depressed.
There will be a rise in temperature.
The udder will appear hot, swollen and sore.
Swellings appear along the abdomen and between the hind
legs.
Because of the pain, the mare may walk stiffly with her hind
legs apart.

Treatment
The vet may take a milk sample in order to identify the
pathogen. Antibiotic treatment is then generally used. In severe
cases, an intramammary antibiotic infusion is used. Some mares

object strongly to this, in which case antibiotics are given systemically (orally or by injection). It is usual practice to strip out (milk) the affected glands frequently.

Blocked teat canal

This can be caused by injury, polyp in the canal or through congenital defect. The vet will unblock/remove the obstruction.

CARE IN THE FIRST FEW DAYS

Vaccination

If the mare was not given a tetanus antitoxin booster six weeks before foaling, or if she 'ran milk' and the colostrum was lost, the foal will require a vaccination against tetanus. Other than this, the vaccination course will start at three months of age.

Handling the foal

The foal must be handled gently but firmly right from the beginning. He must be taught good manners through a consistent routine of disciplined handling. Habits to be discouraged include nipping, rearing and pawing at the ground. It is not desirable to hit the foal on the muzzle as a response to nipping because this could cause him to become head-shy — some advocate pulling sharply on the whiskers to cause discomfort, combined with a verbal reprimand. Rearing and pawing at the ground may also be dealt with verbally, accompanied by a slap on the chest.

The foal should have his feet picked out daily as preparation for future handling by the farrier.

During the first few days, a soft leather foal slip may be put on to the foal. This can be done while an assistant holds the foal — one arm around the chest and one arm around the hindquarters. The foal may be difficult to catch initially — in which case he may have to be cornered through the appropriate positioning of the mare in the loose box.

If mare and foal are to be turned out the foal should be led alongside the mare to the field — this is much safer than allowing him to follow her loose. A soft webbing line attached to the slip and passed behind the foal's quarters is then held in the handler's right hand, level with the foal's loins. The mare is walked on and the foal led at her flank. Pressure from the handler's left hand can be used across the foal's chest to help steady him. Positioned in such a way the foal can (in theory) only move forwards. As the foal becomes accustomed to being led, so the need for the webbing line around the quarters diminishes.

Further handling of youngstock is covered in more detail in the final chapter.

Turning out

If the weather is fine and mare and foal are in good health they may spend some time each day out at grass. At first, they should have no other company but, after three to four days, they may be turned out with other mares with foals of a similar age. Once a group is established (maximum six mares depending on paddock size), further mixing should be avoided. Mares and foals are not normally turned out with geldings.

The paddock

Fencing must be secure and foal-proof — no loose wire or sharp projections. Foals must not be able to roll underneath the fence, especially if there is a ditch the other side. Ideally, boarding extending down to the ground should be used in the first 'nursery' paddock. This principle should also be extended to the void below a water trough. While there must be a constant supply of clean, fresh water, the space below the bottom of a water trough can trap a foal. Furthermore, since foals can drown in ponds and become trapped in ditches these must be safely fenced off.

Grazing must be clean and free of droppings and poisonous plants. All mares must be regularly wormed and foals first wormed at seven days and regularly from then on. The ground must be free of large ruts, and all low branches must be cut well back. Foals generally start to nibble grass within a week and eat grass at about three weeks old. Mares and foals are better on established grass rather than new, and the soil should be tested to ascertain the presence of the correct balance of nutrients.

NUTRITION

The nutrition of the lactating mare and her foal is a subject of great importance — any undernourishment of the mare will reflect in both her condition and in the growth of her foal.

A healthy, newborn Thoroughbred or TB x foal weighing 50 kg should have doubled his birth weight at eight weeks old. Such a foal may weigh approximately 450 kg as a three-year-old and most of this 400 kg gain occurs in the first fifteen months of life. From these facts it can easily be deduced that a high quality, balanced diet is essential.

When calculating feed rations it is necessary to take account of the time of year and whether mare and foal are turned out daily onto clean, good quality pasture. The mare who has foaled early in the season may not have access to the spring grass and the benefits that accompany it. Good quality grass provides an important feed source and its availability will affect the quantities and protein percentages that need to be fed in hay and concentrates.

Foals start to eat concentrates at around seven weeks of age. When feeding a mare and foal at grass a foal creep may be needed — this is a small, corral-type structure which only admits the foal. These can easily be built into the fence line and are useful when feeding to prevent the mare from eating the foal's ration. Some hayracks and mangers used at grass may prove a danger for inquisitive, playful foals who lack any sense of self-preservation — and haynets should never be used when foals are around.

A mineral block may be freely available to both mare and foal.

Considerations in feeding the mare and foal

In addition to those criteria which apply to all feeding, there are some points which require special consideration when feeding a mare and foal:

1) It is essential to provide the foal with the nutrients necessary for healthy growth — the correct balance of all nutrients is essential, with particular attention being paid to the inclusion of sufficient protein, calcium and vitamin D.

2) It is therefore essential to feed the mare in a manner which enables her to produce an adequate supply of good quality milk. (Any dietary deficiency or illness is likely to affect milk production.)

3) It is important to provide a well balanced diet in avoidance of metabolic disorders such as laminitis.

4) In addition to providing the foal with the best possible start in life, it may be necessary to prepare him according to his future competition work. For example, the feeding of a Thoroughbred foal who is to race as a two-year-old will differ from the feeding of a future riding club horse.

5) The feeding of youngsters must ensure provision for maximum growth without leading to fatness — the developing skeleton is not strong enough to carry surplus weight and, indeed, the future soundness of an overweight foal may be adversely affected.

Protein requirements

During the first three months the mare's milk yield is at its highest because the foal is not yet taking much in the form of concentrates. (Foals generally start to eat grass at three weeks and concentrates at around seven weeks.) During this time the mare should be receiving around 14 per cent crude protein in her daily ration. This may be provided in the form of stud cubes, the advantages of which have previously been discussed. Some yards make up their own mixes using compound protein pellets and/or soya bean meal to provide the protein, along with the mare's normal feedstuffs. Good quality lucerne hay or alfalfa helps to provide sufficient protein.

During the first weeks, the foal's protein requirements (16−18 per cent) are met by the mare's milk − for this reason it is essential that she receives her full protein rations. A Thoroughbred mare produces 3−7 litres (5−12 pints) of milk per day soon after foaling, increasing to 9−11 litres (16−19 pints) daily at the peak (which is usually at three months). The mare's milk will not, however, provide all of the protein needed for maximum growth, so supplementary feeding will be necessary. (In the wild the foal would take longer to grow as he would not receive the high protein supplementary feedstuffs.)

In the second three months, the mare's protein requirements lower to 12 per cent, as the foal is usually eating concentrates and grass, depending less on the milk supply for his dietary needs and therefore spending less time sucking. At around three weeks of age the foal may start to suck only every two hours for around five to ten minutes − though all foals are different and will vary in their feeding patterns. As an approximate guide, a foal requires a litre (1.75 pints) of fluid per 10 kg bodyweight per day.

Vitamins and minerals

These must be provided in the correct balance. A mineral lick may be freely available, and supplements added to the diet *as necessary*. (Beware of giving excessive quantities of supplements − overdosing with some substances can be counter-productive, even dangerous.)

As previously mentioned, calcium and vitamin D are important for healthy bone growth. The inclusion of the B vitamin complex maximizes protein utilization, and vitamin A is essential for development and growth.

As with the feeding of all stock, there has to be a balance of art and science. If the owner of a mare and foal is ever in any doubt as to their nutrition, expert guidance should be sought.

12

PREGNANCY FAILURE AND DISORDERS OF THE REPRODUCTIVE SYSTEM

Pregnancy failure may occur for a variety of reasons, not all of which are associated with invasion by pathogens. In this chapter we will examine the stages at which failure may occur, the general causes of failure and the specific disorders which may threaten or prevent a successful pregnancy.

PREGNANCY FAILURE

Pregnancy failure may be referred to by the more specific terms resorption (the embryo being resorbed in early pregnancy), foetal death and abortion. Abortion can be defined as the expulsion of the uterine contents before term. The mare usually appears none the worse for having aborted and, in a grass-kept mare, it may go unnoticed: the products of the abortion may not be clearly visible, especially if scavenging animals have found them first.

Failure at different stages

While a pregnancy may fail at any stage the underlying reasons and immediate consequences will vary:

During Week 1. Although impossible to detect definitively, it may be assumed that some fertilized eggs do not reach the uterus, but die within the fallopian tube.

During Week 2. The fertilized egg normally reaches the uterus five days after ovulation. Inflammation of the endometrium or regression of the corpus luteum will lead to an inhospitable environment within the uterus, making it impossible for the conceptus to develop.

During Weeks 3—5. The corpus luteum should still be producing progesterone, which stimulates the uterus into preparation for pregnancy and continues to suppress oestrus. If, as a result of a hormonal imbalance, the corpus luteum regresses, the mare may return to oestrus. The fluids of the embryo are resorbed into the mare's bloodstream, and any remnants of the conceptus will probably be passed out unnoticed during the next heat.

During Weeks 6—20. Should the foetus die, the fluid may be resorbed. The solid tissue will degenerate as a result of the release of cell enzymes, and may be expelled during the subsequent heat. If the foetus is lost after endometrial cup formation, PSMG still persists until around day 120, and the mare does not return to oestrus before this time.

From Week 21—Term. After four months, the foetus has a recognizable skeleton. In the event of death, (for example of one of a pair of twins) dehydration and, to a degree, resorbtion will occur, leaving a 'mummified' foetus. This may be present at the birth of the surviving twin. Mummification rarely occurs when a single foetus dies because foetal death normally causes abortion.

Causes of pregnancy failure

1) Twinning. This has already been discussed in the chapter Diagnosis of Pregnancy.

2) Foetal abnormalities. These include deformities resulting from genetic abnormalities (for example, hydrocephalus — fluid on the brain).

3) Twisted umbilical cord. Should a twist be severe enough to impede circulation (and therefore reduce the oxygen supply), foetal death will occur.

4) Reduced placental function. Chronic changes or damage to the placenta (for example, placentitis) can lead to fatal malnutrition of the foetus. Fungal placentitis usually affects the cervical pole, causing gross thickening. It results in late abortion, or the birth of an impoverished foal.

5) Infection. This may be bacterial, viral or fungal (see above). Especially in mares suffering from pneumovagina (discussed in the next section) bacteria may enter the vagina, causing inflammation which may spread to the uterus.
Bacterial infection can also lead to septicaemia, which may be transferred to the foetus via the umbilical cord. Specific bacterial and viral infections are discussed in the following section.

DISORDERS OF THE REPRODUCTIVE SYSTEM

Prolonged dioestrus

This non-infectious condition is relatively common and may affect both breeding stock and maiden mares. The cause is that the corpus luteum persists, sometimes for up to three months, so the mare fails to return to oestrus.

The vet may take a blood sample to aid diagnosis: cases of persistent corpus luteum have levels of progesterone higher than 4 mg/ml.

Treatment is by injections of the hormone prostaglandin,

which will cause luteolysis — regression of the corpus luteum. This will allow the oestrus cycle to resume.

Pneumovagina (vaginal windsucking)

Ideally, the top of the vulval opening should be level with the pubic bone, in order that the vestibular seal is functional. Faulty conformation of the vulval opening (or damage and overstretching of the vulvo-vaginal constriction) allows air to be drawn into the vagina. As the vulva is so close to the rectum, this air is often contaminated. Bacterial infection may occur which, if drawn forward towards the cervix and on into the uterus, may lead to endometritis. This inflammation of the uterus lining prevents pregnancy. Apart from infertility, other signs may include a noise made by the mare whilst walking.

If, as a result of pneumovagina, infertility is a problem, the veterinary surgeon will probably advise the use of Caslick's operation, in which the vulval lips are cut and then partially stitched together (sealed down to the level of the pubic bone) in order to reduce the opening.

Once the mare does become pregnant, an episiotomy will have to be performed when foaling is imminent.

Contagious equine metritis (CEM)

This highly contagious bacterial venereal disease is a serious threat to breeding stock as it causes greatly reduced rates of conception. It was first diagnosed and identified in Great Britain in 1977, before which it was unheard of.
Because of its effect on fertility, CEM is now a notifiable disease: where it is suspected, a report must be made to the Local Divisional Veterinary Officer at the Ministry for Agriculture, Fisheries and Farming and, upon confirmation of a diagnosis in Great Britain, The Thoroughbred Breeders' Association at Newmarket must be notified.

Causes
The bacterial CEM organism is found mainly around the area of the clitoris, cervix and urethra in the mare. Stallions may carry the organisms over the surface of the penis and in the

urethral fossa, but show no clinical signs of the disease.

The organisms may pass from mare to stallion or vice versa during covering. If the mare is infected around the clitoral and cervical area, covering or manual examination may push the organisms forward into the uterus, causing endometritis. Endometritis leads to infertility as it causes the early resorption of the hormone-producing corpus luteum. This results in the mare repeatedly coming back into season at shorter intervals than normal — 'short-cycling'.

Signs
In some mares, there may be no obvious signs: others show a profuse, purulent discharge from the vulva.

Treatment
If CEM is confirmed in mares prior to mating, infected mares must be isolated and treated as advised by the vet. All owners who have booked mares to any stallion, or whose mares have left the stud, must be notified.

If CEM is confirmed after mating the above procedure must again be followed. In addition, the stallion should cease covering mares and must be swabbed and treated as advised by the vet. Covering should not resume until the stallion has been treated and has had three negative sets of swabs. The first swab should not be taken until at least seven days after the completion of treatment and subsequent swabs should be taken at intervals of not less than two days. The stallion should then be test-mated to at least three mares, who are swabbed afterwards. All mares implicated must be checked — this may include blood tests.

Rhinopneumonitis (EHV-1)

This viral disease is caused by the equine herpes virus 1 (EHV-1) and occasionally by equine herpes virus 4 (EHV-4). When breeding stock are affected, the condition is often referred to as viral abortion.

Primarily, the virus causes respiratory disease and, in the case of EHV-1, paralysis. The source of infection may be nasal discharge, aborted foetuses, placentas and/or associated fluids: this disease is not spread through covering. It is a serious threat when broodmares are affected, because of the highly infectious nature of the disease and the high rate of abortion caused as a result of infection. Once abortion has occurred, the future breeding capacity of the mare will not normally be impaired.

Signs

Signs of the respiratory form of the disease include mild fever, coughing and nasal discharge. These signs are usually apparent in weaned foals and yearlings. Older animals may contract the respiratory disease without showing any signs. The respiratory form occurs most frequently in the autumn and winter.

Effects

Pregnant mares may abort from two weeks to several months after infection. Abortion may occur as early as four months into pregnancy, but is more common between the eighth and eleventh months, with the eighth and ninth months showing the highest abortion incidence. The aborted foetus is often delivered in the membranes — alternatively the membranes are expelled promptly.

In the event of an infected mare going to full term, the foal may be born alive but will be very ill and highly infective. He will be weak, jaundiced, and may have difficulty breathing. Very often a foal so ill will die within a few days. There is a very high incidence of stillborn foals born to infected mares who actually go full term.

Diagnosis and control

In the event of abortion occurring, the foetus and membranes should be sent for post mortem examination, so that the cause can be accurately confirmed. If EHV-1 is the confirmed cause of the abortion, The Thoroughbred Breeders' Association must be notified. A post mortem examination should also be carried out on dead foals.

All other products of the abortion (the fluids) must be burned, along with the bedding. The bedding and other matter must be sprayed liberally with disinfectant in the box and left for forty-eight hours prior to burning, in order to minimize the risk of infected particles being blown around when moving the bedding material from the stable to the incineration area. The person in charge of these procedures should not be in charge of other pregnant animals.

The loose box must then be steam cleaned and soaked in a disinfectant, preferably one from the iodophor range, such as Pevidene or Wescodyne. If this cleansing is not carried out diligently the virus may live for several weeks within the structure.

If a mare aborts in the field as a result of EHV-1, the other mares should be kept as a group or divided into smaller groups, but should not be mixed with other pregnant stock.

The affected mare must remain in isolation for at least eight weeks. She may be allowed to leave the stud one month from the date upon which the last mare aborted at that stud, but she must not come into contact with any in-foal mares for at least two months.

Precautionary measures

Any mare going to stud to foal should arrive at least one month before the due foaling date, and go into small-group isolation. If a mare in late pregnancy arrives at stud to foal from abroad or from the sale ring, she must be isolated individually until after foaling. Pregnant mares should not travel with other stock, in particular with recently aborted mares or youngstock. If ever a foster mother is to be used on stud, she should be individually isolated until it has been confirmed that her own foal did not die from rhinopneumonitis. If EHV-1 is confirmed, the appropriate breeders' association should be informed.

Vaccination

There is, at present, one vaccine available to help prevent multiple abortions. Pneumabort K contains inactivated whole virus with adjuvant to enhance the immune response. The manufacturers recommend a vaccination programme for pregnant mares and all in-contact animals.

Pregnant mares are vaccinated every two months from the fifth month of pregnancy. Barren mares and teasers should have an annual booster. The Horserace Betting Levy Board has prepared a *Code of Practice for the United Kingdom and Ireland* with regard to all aspects of equine herpes virus 1.

Coital exanthema

This venereal disease is caused by the equine herpes virus 3 (EHV-3) and is often spread by coitus. It is, however, difficult to be totally accurate as to the initial cause, especially as it has been found in non-breeding stock.

Both mares and stallions develop ulcer-type genital sores which rupture, leaving infective open sores. Normally, infertility only results insofar as the soreness of the genitalia prevents coitus.

Treatment entails rest from sexual activities and the application of antibiotic substances to affected areas in order to prevent secondary bacterial infection. The condition normally clears up after seven to fourteen days.

Equine viral arteritis (EVA)

This highly infectious viral disease had never been seen in Great Britain until the spring of 1993, when an outbreak originated from an infected Polish stallion who had been imported in September 1992. At the time of writing the UK has no power to blanket test all imported horses or to impose import restrictions, as this would contravene EC ruling. The Ministry of Agriculture can only carry out random checks on imported animals, a factor which may contribute to further outbreaks of the disease.

EVA is transmitted via droplet infection from the nose and mouth of infected animals and via the semen of stallions who have become 'carriers' following infection.

Signs
Over 50 per cent of pregnant mares will abort, without necessarily showing other clinical signs. Otherwise:
High temperature − 105 °F (40.5 °C) for one to five days.
Loss of appetite.

Depression.
Conjunctivitis (also known as 'pink eye').
Nasal discharge and congestion.
Swollen glands in the throat region.
Oedema on head, throat and legs.
Some horses experience diarrhoea.
In foals, other signs include respiratory distress, pneumonia and colic, while mares may develop swelling of the mammary glands and stallions develop swelling of the scrotum. Death may occur, especially in foals.

After-effects
Although horses who survive an attack require a long period of rest and rehabilitation, the aftermath of the disease varies between the sexes. Of those stallions who recover, about a third become carriers of the disease (known as 'shedders'). Long-term shedders pass the virus on through semen only, but this renders them unusable for stud work. Infected mares recover from the worst signs relatively quickly and stop shedding after some three weeks, at which stage they have developed natural immunity.

Treatment and vaccination
Beyond supportive therapy to alleviate the signs there is no treatment. A vaccine is available but it will produce a lifelong antibody which may render valuable bloodstock non-exportable under current export regulations.

Serological examination will detect the presence of antibodies in blood, serum or other body fluid. The antibodies are developed in response to infection or vaccination. A horse tested sero-negative shows negative results — antibodies to a specific micro-organism (in this case the equine arteritis virus), are not present in the serum. A seropositive result indicates that the EVA antibodies are present but it is not possible to determine whether this is as a result of infection or vaccination.

If a horse is seropositive, a second sample is taken two weeks later. If the second titre (antibody level) is the same or lower, it is concluded that infection or vaccination occurred sometime previously and the horse poses no risk. If the second sample shows a higher titre, active infection is implicated. In the case of seropositive stallions, semen can be collected and cultured to detect a shedder.

If a seronegative stallion is to be vaccinated, it is wise to obtain documentation of his seronegative status prior to vaccination to prove that his seroconversion is a result of vaccination and not infection. The Horserace Betting Levy Board's Code of Practice recommends that horses — particularly those likely to be traded overseas — should be serologically tested immediately prior to vaccination. This will prove that the horse has not been infected and that subsequent seropositive status is as a result of vaccination.

Prevention

Subject to veterinary advice, any horse imported from a country where EVA is known or suspected to occur should immediately be isolated and remain in isolation for a minimum of twenty-one days. Blood samples should be taken on or before arrival, and again after at least fourteen days. This is particularly important with stallions intended for mating or semen collection for AI.

All stallions used for AI should be tested negative for EVA at the beginning of every breeding season, and any other stallions standing at the same stud should also be tested negative.

All consignments of semen imported into the country should be accompanied by a certificate verifying that the semen is derived from a negative-tested stallion.

Recommended action to be taken if EVA is confirmed or suspected is as follows:

1) Stop all movement on and off the premises.

2) Cease all covering.

3) Isolate the infected animals and all in-contact animals.

4) All animals on the stud will need to be serologically tested.

5) Isolate all horses according to their infectious status:
   Group 1 — clinical cases with all seropositive horses.

Group 2 — healthy seronegative in-contacts.
Group 3 — all healthy seronegatives.
If practical, use separate staff for each group.

6) Aborted foetuses, membranes, nasopharyngeal swabs, semen, and serum can be tested for confirmation.

7) Serological tests should be repeated after fourteen days, and groups reassessed and isolated accordingly.

8) Infected mares, geldings and foals on the stud should stay in isolation for at least one month after they become sero-positive. Further tests will have to be carried out.

9) All collection of semen for AI should cease until the stallion has been isolated for at least two weeks and the results of the second blood test are known.

10) The disease must be reported to the relevant breeders' association and owners whose horses are at the stud or due to arrive must be informed.

11) All stables and horseboxes must be thoroughly cleaned and disinfected.

12) Veterinary advice must be followed at all times.

13

AILMENTS OF THE FOAL

In this chapter we will look at both non-infectious and infectious conditions which are either specific to foals, or of particular concern when they appear in youngstock.

NON-INFECTIOUS CONDITIONS OF THE NEWBORN FOAL

Prematurity and dysmaturity

A foal born between 300 and 320 days gestation is said to be premature. A dysmature foal is one born at or about full term but showing signs of prematurity/immaturity, probably as a result of malnourishment or birth trauma.

Signs
Weak and underweight.
Delay in standing after birth.
Difficulty in sucking.
Tendency to develop colic and retain meconium.
Susceptibility to infection.
Silky, fine coat; floppy ears.
Hyperextension (down on fetlocks).
Red/orange-coloured tongue.

Treatment
Should be as directed by vet — antibiotic cover is usual. Keep the foal warm.

Check the level of colostral immunity (IgG) and correct if necessary with bottle-fed colostrum.

Neonatal maladjustment syndrome

This is also referred to as 'barking', 'wandering', 'dummy' or 'convulsive' foal.

Signs
Within twenty-four hours of birth the foal loses the ability to suck. He appears blind, staggering around the box. He may have convulsions and go into a coma.

Causes
Not known, but trauma and/or oxygen starvation are suspected.

Treatment
Consult the vet. Very careful nursing will be necessary. The foal may need to be fed through a stomach tube if unable to suck.

Entropion

Signs
In-turned eyelid and weepy eye.

Causes
The eyelid is turned inward, causing the lashes to irritate the surface of the cornea and produce weeping. This condition is more likely to affect newborn foals in poor bodily condition (through reduced fat pad behind eyeball).

Treatment
If not treated, this irritation will lead to keratitis, ulceration, blindness or complete loss of the eye. Therefore, call the vet immediately. The vet will stitch the eyelid back for a period of time.

Haemolytic disease

Because of the destruction of the red blood cells by incompatible proteins in the colostrum the newborn foal (first three to four days) becomes jaundiced and anaemic. If severe or undetected, this condition is fatal.

Signs
Pale or yellow membranes.
Loss of appetite — will not suck.
Dull, lethargic appearance.
Increased respiratory rate; panting.
Red urine caused by broken down products of the red blood cells passing into the urine.

Causes
Incompatibility between the blood of the sire and that of the dam. This leads to the formation of antibodies which the foal ingests via the colostrum. The antibodies break down the foal's red blood cells.

Treatment
A blood transfusion will be necessary. The mare's colostrum should be stripped and discarded. (This should be done automatically when she has her next foal — the foaling must be supervised and the foal muzzled to prevent any colostrum being taken.) Feed the foal replacement colostrum and only allow him to suck from the mare once her colostrum is finished.

If a mare is known to have produced a jaundiced foal, a blood test may be taken during the next pregnancy to test the compatibility of her blood against that of the sire, or to detect the build up of haemolytic antibodies in her circulation.

Meconium retention

Signs
Straining, squatting and rolling; signs of discomfort associated with colic.

Causes
Not fully known, since colostrum normally stimulates bowel

movement.

Treatment

The vet or an experienced stud attendant should administer a soapy enema or liquid paraffin. Meconium in the rectum immediately behind the anus may be removed with a lubricated, sterile gloved finger.

Diarrhoea

Diarrhoea is not an actual disease, but a sign of internal upset. It must be taken seriously, because it can quickly cause dehydration and because bacterial infection may be implicated.

Signs

Scouring: the area around the tail becomes badly soiled. The foal may flick his tail repeatedly from side to side, causing further irritation.

Causes

Foal heat in the mare. Ailments such as metritis or laminitis affecting the mare can also affect the milk.

Treatment

If caused by the foal heat, scouring should cease when the heat ends. The foal may need fluid therapy to maintain levels of body salts.

If other causes are indicated, isolate and call the vet. Faecal samples may be taken for analysis to determine the cause. The vet will administer treatment for infection as necessary. Diarrhoea is commonly caused by the bacteria E.coli, but can also be caused by Salmonella spp, in which case there is also septicaemia and the risk of handlers becoming infected. Therefore, high standards of hygiene must be maintained.

NON-INFECTOUS CONDITIONS OF THE OLDER FOAL

Hyperflexion (contracted tendons)

Signs
Fore- and/or hind limbs are straight or 'knuckling over'. The foal may 'knuckle over' gradually or suddenly.

Cause
The definite cause is not known, but it is possible that a deficiency of calcium, phosphorous, vitamins A and D and/or hereditary factors contribute.

Treatment
Physical therapy will help but if there is constant 'knuckling over,' more aggressive action has to be taken. This may entail correction of deviations using plaster casts and/or surgical shoes. Care is needed with plaster casts, as sores develop rapidly. Therefore casts must be changed regularly — gutter casts changed twice daily would be the preferred method.

Epiphysitis

This inflammation of the growth plates may affect yearlings and two-year-olds as well as foals. Its location will vary according to which joints are affected.

Signs
The joints of the limbs may appear inflamed, there may be lameness and the foal may walk with a stiffened gait. Severe cases may result in angular limb deformity.

Causes
May be one or a combination of the following:

Jarring and concussion on hard ground — a particular risk with heavy, overweight foals.
Imbalance between calcium and phosphorous in the diet.
Sudden parasitic burden causing a dietary imbalance.
Some horses have a hereditary predisposition to the condition.

The condition may be complicated by disorders of the muscles, tendons and/or ligaments.

Treatment
Abnormalities must be corrected early enough to prevent permanent skeletal deformity. Consult the vet, who will advise on the correction of the limb posture. Special corrective boots may be fitted and later, specially made up shoes.

It is important to ensure that the diet is correctly balanced — supplement vitamins A and D, and calcium.

Do not leave the foal out permanently on hard pasture. Box rest may be necessary.

Osteochondrosis

Osteochondrosis is a disturbance of conversion of cartilage to bone during bone formation, resulting in a persistence of cartilage in the area under the articular cartilage.

There may be separation of a fragment of articular cartilage and underlying bone. Such a fragment, referred to as a joint mouse, causes pain and varying degrees of lameness. Where such a fragment is present, the condition is referred to as osteochondrosis dissecans (OCD).

Signs
It is most commonly seen in one or both stifle or hock joints of young horses, where it may show as inflammation, stiffness and/or lameness. Diagnosis will be confirmed by x-ray.

Causes
These are not known for certain, but there could be a hereditary predisposition.
Heavy topped, rapidly growing foals are more susceptible, and nutritional imbalances — especially those involving calcium and phosphorous, have been implicated.

Treatment
If inflammation is present, anti-inflammatory therapy will be required. Because of the need for analgesia, phenylbutazone may be prescribed. Where there is a fragment present, surgical removal may be necessary to avoid the development of

degenerative joint disease (DJD).

The long-term outlook depends upon whether or not DJD develops.

INFECTIOUS DISEASES OF THE YOUNG FOAL

Joint-ill

This very dangerous infection is sometimes referred to as 'navel-ill' and 'infectious arthritis'. It affects foals from a few days old to around five months. It must be classified as a medical emergency and, in the case of any lame foal, this condition must be ruled out first.

Signs
The foal will be dull, with a fever, and off-suck. The first indication of the condition may be that the mare has a full bag.
One or more hot, swollen joints.
The navel may be swollen and painful.
There may be abscessing along the umbilical vein. This is not always visible from the outside — a scan may be needed.
Deterioration may lead to death.

Causes
Lack of colostrum.
Bacterial infection at the navel leading to infection of cartilage and bone within joints. (Associated with poor hygiene after foaling.)

Treatment
Call the vet, who will administer antibiotics as necessary and flush the affected joint(s).
If there is an abscess at the navel this will have to be removed surgically, otherwise it may spread to other joints.

Rhinopneumonitis (EHV-1)

Foals are particularly susceptible to this highly infectious condition of the upper respiratory tract. Rhinopneumonitis has been discussed in depth in the chapter Pregnancy Failure and Disorders of the Reproductive System. In foals, the signs are nasal discharge, cough and depression.

If a foal succumbs, consult the vet, isolate, and adhere to rules of sick nursing.

Pneumonia

This serious condition generally affects foals in the first four to eight weeks of life.

Signs
The foal appears very apathetic and poorly.
Rapid, laboured breathing; coughing.
High temperature — the temperature rises progressively and, if not treated, death results.

Causes
The cause may be viral or bacterial.
Viral agents are: herpes virus; adenovirus; rhinovirus.
Bacterial agents are: Streptococci; Staphlococci; E. coli; Pasteurella; Klebsiella.

Treatment
Call the vet as soon as the foal shows any signs of ill-health such as a rise in temperature. In cases of pneumonia the vet will administer the appropriate antibiotics — this treatment may have to be carried on for approximately two weeks after the foal appears to have recovered.

The foal must be kept warm, in a well ventilated stable.

Neonatal septicaemia

This bacterial infection generally affects foals within the first week, and is usually fatal.

Signs
The various bacteria which cause neonatal septicaemia localize in different body organs, so different signs predominate accordingly. An example of one form of septicaemia is 'sleepy foal

disease' — the organism Actinobacillus equuli localizes in the kidney, causing nephritis.

In general, signs to watch out for are as follows:

Initially, the foal appears restless and nervous — he either sucks for very short periods or is off-suck.
After a few hours the foal appears to lose strength and lies down.
Rapid breathing.
Either high temperature or hypothermia.
Membranes may become dark red.
Sleepiness.
Sometimes mild colic and diarrhoea.
Convulsions and unconsciousness, followed by death.

Causes
Usually, bacterial infection by (in approximate order of incidence): Streptococcus zooepidemicus; E. coli; Actinobacillus equuli; Staph. aureus; Salmonella spp; Klebsiella spp. Sometimes, a consequence of viral infection (EHV-1).

Treatment
Call vet very promptly: intensive antibiotic therapy may help, but the condition often causes death within twenty-four hours.

Prevention
Is through provision of adequate colostral immunity and good hygiene in the foaling box.

Rotavirus

This condition affects foals up to five months old but typically occurs during the first two months.

Signs
Profuse, foul-smelling diarrhoea, and dehydration.
Mild colic.
Weight loss.
Scalding of perineum.
Depression.
Fever.

Cause
A virus which damages the lining of the intestinal tract.

Treatment
Confirmation is through testing of faecal sample. Fluid therapy is used to prevent dehydration, and antibiotics will be given if secondary bacterial infection occurs. Grease (zinc oxide or Vaseline) may be applied to the perineum to prevent/alleviate scalding.

Gastroduodenal ulceration

This is often a sequel to rotavirus infection and is predisposed by stress (caused by travel, surgery, weaning, mixing groups etc.) and other infections.

Signs
Increased salivation.
Grinding teeth.
Mild colic — looking at flanks, rolling, pawing.
Off-suck.
Variable temperature.

Treatment
The vet will administer anti-ulcer medication.

14

GENERAL MANAGEMENT

The way in which a horse is handled has a profound effect on his overall behaviour — the horse soon learns what is allowable and acceptable behaviour and what is not. These definitions will only become clearly apparent through confident handling and consistent attention to the 'rights and wrongs' of behaviour. In this section of the book we shall discuss the general handling of stallions and young horses. (It is assumed that the reader already has significant experience in the day-to-day handling of horses.)

STALLION MANAGEMENT

Because of the high value of many stallions, they are managed in a way that entails minimum risk. This may present problems if a stallion spends long periods of time stabled, resulting in boredom. Since stallions are normally very strong and physically fit, boredom can lead to antisocial behaviour. It is therefore imperative that they are handled in a confident and competent manner, with insistence on a high degree of discipline and good manners. (The early management of a colt influences his behaviour in later life. Vices such as biting, rearing and striking out should be corrected early on since, once established, they are difficult and dangerous to contend with.)

When working with any stallion, whether a large Thorough-bred or small Welsh Mountain pony, the handler must try to gain his respect through skilful and firm treatment. However, the degree of 'firm' treatment required will depend upon the stallion's nature — there are those who are extremely placid and no more difficult to look after than any mare or gelding. Also, remember that a stallion is a strong, intelligent animal deserving respect himself — aggressive treatment from the handler will probably result in a battle of wills and a breakdown in the relationship.

Whatever the stallion's temperament, consistent discipline, sufficient exercise and reducing boredom will contribute greatly to a better-mannered horse.

Nutrition

The stallion will require a high quality diet during the breeding season. This should provide 9.6 per cent protein for mature stallions; 11–13 per cent for young stallions. Stud cubes are specially prepared to provide this protein, as well as a balance of all vitamins and minerals to help ensure maximum fertility.

A stallion must be fed only good quality grass or hay; the provision of ad lib hay also helps to prevent boredom in a permanently stabled stallion.

General care

The stallion must undergo all of the normal health checks — teeth rasping, worming, vaccination and shoeing. If the stallion is not competing or hacking out regularly he may not require shoeing, but his feet will need to be kept well trimmed.

He should be strapped regularly to improve his appearance and to ensure a clean, healthy coat. It may be necessary to tie him up using pillar reins if he is prone to biting. Care must be taken if there is a risk of him kicking out — one foreleg may need to be strapped up.

In the stable

While some studs have a separate stallion yard, other stallion boxes are in mixed yards. In the latter case, a stallion should be able to see things going on in the yard but be away from any mares.

The larger his stable, the better. The stable must be completely free from projections and there must be no bars which may trap a foot (for instance, on a window, hayrack or bucket racks). The bottom stable door must be very sturdy and high enough to deter the stallion from putting his forefeet over it, or jumping out. In some cases, a top grille may be necessary to prevent biting. The stable must be well ventilated and clean.

In the paddock

Every stud has a different policy regarding the turning out of each individual stallion, depending on his temperament and value. Ideally, the stallion ought to have a period each day spent out at grass to relax, exercise himself and reduce boredom.

Paddock fencing must be very high — at least 2 m (6 ft 6 in) and preferably in the form of a screen, which will help prevent the stallion from seeing passing horses and attempting to jump out. Electric fencing either along the top or approximately 20 cm (8 in) in from the fence provides extra security. The corners of the fence must be rounded, as in normal paddocks.

During the winter some stallions are turned out with the mares — note that a gelding in the same field may be attacked by the stallion.

Native pony stallions may run with the mares and cover them naturally.

Exercise

Depending upon the number of mares in 'the book' a covering season may be strenuous, and a stallion will need to be physically fit to cope with its demands. Exercise is therefore very important — it reduces boredom, improves cardiovascular function and promotes fitness.

It is also important that the stallion does not associate coming out of the stable solely with covering a mare. He must also come out of the stable for exercise, which may be in the form of turning out, ridden or in-hand exercise. (When leading a stallion, the handler must not take chances. A Chifney bit is often used to help contain high spirits. The handler should wear gloves and a crash hat, and carry a stick to aid discipline.)

If a stallion is to be exercised under saddle, the rider must be competent. Whether to exercise alone or in a group will depend on the stallion's manners, temperament and the time of year (whether it is during the breeding season or not).

Lungeing and loose schooling are useful methods of exercise if undertaken by an experienced handler.

HANDLING YOUNGSTOCK

A young horse has a lot to cope with in the first years of life — he has many experiences to undergo, some enjoyable, some not. His first contact with humans must be a pleasant experience, worthy of his trust. He should then learn about being handled, led and groomed.

Experiences that are not enjoyable include weaning and castration — through careful planning, these should be made as stress-free as possible.

Early handling of the foal

The sooner a newborn foal is handled, the better. Initially, the foal may be very shy and hesitant about being handled. In some cases it helps to mask the human smell by rubbing your hands in the mare's coat or bedding.

Foals are usually calmer if they are kept near the mare, so drive a foal between the mare and a wall to catch him. Approach quietly and allow the foal to sniff at your outstretched hand. Talking quietly to him, gradually edge him into a corner and rub his neck and withers.

When possible, put on a foal slip. Restrain the foal with one arm around his chest and one below the rump — if you pull on the foal slip the foal will inevitably go over backwards. The foal may try to struggle but you should remain firm but quiet with him. Keeping hold of the slip, walk towards the dam. You may make him stop before he reaches her and give him a lot of praise — these are the foundations upon which all of his future lessons are based. Care must be taken with the foal slip — if too loose it may be snagged on something or the foal may put a foot through it.

For the first two to three days of his life the foal will be constrained within a large, airy loosebox with his dam. During these few days you will be able to make friends with him so that, by the time he goes out into the paddock for the first time, he is more likely to trust you — it is virtually impossible to approach and handle a very shy foal out in the field.

Each day, time should be spent in the box before turning out. The foal can be led for a few steps and praised well. You may need someone to help you as you gently run your hands over the foal's body and down his legs to accustom him to being touched. As mentioned above, the best way to restrain a foal is to have one arm around his chest and one around his hindquarters. In the case of a lively colt foal this may require a strong person. Once the foal is accustomed to the feel of your hands over his body you can introduce a soft brush. Do not 'overdo' the grooming, particularly if the foal is spending a lot of time out at grass, as he requires the natural grease in his coat for warmth.

After a week or so, quietly pick up a foot and get the foal used to standing like this for a short time. Remember to do this equally with all four legs — some youngsters will happily pick up their near fore but can be very uncooperative about the other three! Pick out the feet and 'tap' the foot walls with the hoofpick to accustom the foal to the feel and sound.

Teach the foal to tie up by passing the lead rope through the tying ring and holding the end of the rope. An assistant should be present to apply pressure behind the foal's hindquarters should he try to pull backwards. As the rope is not actually tied, the foal is not in danger of hurting himself by struggling.

Once he stands without resistance, the rope may be tied to a weak link, using a quick-release knot. Tie up for short spells only, gradually increasing the duration as the foal becomes accustomed to it. Never leave a foal tied up unattended — should he pull back he could damage his neck and/or lose his footing and fall.

The first summer

If the weather is very hot and flies are a problem, both mare and foal will be happier stabled during the day and turned out at night.

The foal's handling can be increased, to include practice in leading, as required in the show ring. The mare is led slightly in front of the foal at walk and trot. The foal must follow behind, either trotting or cantering — preferably not bucking wildly! The pair may then be led to stand patiently 'in line'. Once in line at a show, the mare and foal will be pulled out and the mare normally trotted up first. Practise this at home — the foal's handler should turn him away and possibly give him a handful of grass as a distraction. This exercise is then repeated with the mare as the foal is trotted up. This procedure should be well established before the first show in order to avoid a possibly embarrassing lack of co-operation from the foal at the vital moment!

Both mare and foal should be taught to stand up properly — a showing stick should be carried as an aid and can be used to tap the legs gently if necessary.

When the foal reaches the age of roughly three months his feet may require rasping by the farrier. Any uneven wear of the feet will cause uneven pressure to be exerted on the joints and tendons, and this may lead to faulty action and possible joint damage. However, an experienced farrier may be able to correct any uneven wear and faulty action at an early age.

While the foal is having his feet attended to, he should be quietly restrained by the method described earlier. This is much better than have a pulling match on the end of a rope, which may result in the foal rearing and possibly hurting and frightening himself.

When the foal is out at grass it is an excellent idea to catch him at the same time as his dam is caught and to give him a reward — this is an important lesson which will pay dividends in the future. Rewards need not be in the form of titbits, which might encourage nipping: verbal praise, patting and a handful of grass should suffice.

Practise leading the foal from both the near and off sides and, at all times, insist on obedience and discipline — this should be encouraged through reward and correction as a result of constantly fair but firm handling. Fillies are generally easier to handle than colts — indeed some colts are very stallion-like in their antics and need particularly firm handling.

The first winter

During the first winter the general handling can continue. The foal may live out provided there is very good shelter with plenty of bedding — youngstock should not be allowed to get cold and wet. Obviously, their breeding and the severity of the weather will help to determine whether the mare and foal should live out or not.

Although the foal will probably not need to wear one it may be useful to introduce a rug and surcingle. However, it is better not to 'mollycoddle' the foal — provided he is kept warm and dry, his coat and natural greases should be adequate.

Care must be taken when turning out in icy weather — poached ground which has frozen solid in rough, uneven potholes poses a great danger to young limbs. Even walking across this rough ground can lead to strained tendons, and should therefore be avoided.

During the winter the mare's milk will start to decrease, so the foal must be gaining sufficient nutriment through good quality hay and concentrates.

Transporting mare and foal

When transporting a mare and her foal a lorry is more satisfactory than a trailer. If a trailer has to be used, the centre partition should be removed. Ensure that the ramp is not too steep — it is essential that it is non-slippery and firm. Bed the floor well with straw or shavings. There must be nothing in which a foal could become trapped (such as low haynets).

The mare and foal should be led up the ramp together and the mare tied up in the normal manner. Ideally, an assistant should travel, holding and reassuring the foal. (This is illegal when travelling in a trailer.) If there is no assistant, the foal should stand loose, preferably in a (visibly) partitioned-off section in front of the mare.

Once the lorry is loaded, the ramp should be closed very quietly and carefully. The lorry must be driven very carefully and slowly to avoid frightening the foal.

Each stud develops its own preferred method of transporting mares and foals, and their advice should be sought when the mare and her offspring are to be brought home from stud.

CASTRATION

Colts are castrated for the following reasons:

To enable male and female animals to be kept together.

It keeps the strains pure, as random breeding is prevented.

Entire horses can be difficult to handle, particularly in the presence of mares in season.

The exact age at which a colt is castrated is dependent upon his physical development. Some colts are slightly 'weedy' and benefit from a later castration. However; well developed colts can be castrated at around five months. The younger the colt when castration is performed, the better, as it is less traumatic and more humane. The unweaned colt can also return to his dam after the operation for reassurance.

Generally speaking, colts are castrated at around one year old that is, in the early spring of the year following their birth. Some late foals may need to be left until the autumn. If a colt is thought to have breeding potential he may be left entire until approximately two years old — if he does not reach his potential he may then be castrated.

Methods of castration

Castration must be carried out in cool, fine weather — preferably not in the heat of summer when the flies will cause irritation and possible infection of the wound. Prior to castration the colt should not eat too much, keeping his intestines free of excess bulk.

The operation consists of opening and cutting the scrotum and, using a specially designed instrument called an 'emasculator', crushing the spermatic cord. The cord is cut below the crushed part. The crushing helps to prevent bleeding from the spermatic artery.

The following terminology is used to describe the precise methods of castration:

Open/closed. This refers to whether the tunic and scrotal skin are left open (leaving a potential risk of herniation of intestines) or closed, whereby the tunic and scrotal skin are ligatured/closed off. The former method is preferred under 'field conditions' (not in an operating theatre), as it allows free drainage and reduces the risk of infection or swelling.

Standing. Often used in Thoroughbred racehorses: the horse is tranquillized and castrated using the open method under local anaesthetic.

Anaethetised. Under 'field conditions' the horse is anaethetized with a mixture of injectable anaesthetic agents and muscle relaxants.

Under theatre conditions. The choice method from the point of view of sterility and the only method for crytorchids, in whom it may be necessary to explore the inguinal canal or abdomen to find the retained testicle(s).

Aftercare

If he is unweaned and the weather is agreeable, the colt may remain in the field with his dam. If he has already been weaned he will need to be kept quiet in a well disinfected stable for a day or two.

The vet will check him during the following twenty-four hours for post-operative bleeding, infection or hernia.

WEANING

There are no hard-and-fast rules concerning the timing and method of weaning; each individual stud has its own policy. Factors which may affect the time of weaning include:

Date of foaling — six months is a popular age for weaning.

Nutritional independence — the foal must be eating hay and concentrates well.

Whether the mare is in foal or not. If she is in foal then her offspring may need to be weaned around September as the grass loses its nutritional value thereafter. The date of covering will affect the date of weaning — the unborn foetus is developing rapidly during the last three months of pregnancy and making great demands of the mare, therefore the foal must be weaned by eight months. If the mare is not in foal, it will not be necessary to wean so early.

If the foal is 'heavy topped' or carrying too much condition, and therefore at risk of developing skeletal disorders (epiphysitis or OCD), it is preferable to wean early.

Methods of weaning

Weaning can be carried out either at grass or in the stable.

At grass

A group of mares and foals may be turned out together for a reasonably long period to ensure that they know each other well and feel secure. Remove one mare, preferably the dam of the most forward, independent foal, to a distant, well fenced field with a placid companion. Over a period of days, remove the other mares, one at a time, leaving one mare as guardian. She too can be removed when the time is right.

This method of weaning reduces the stress for both mare and foal.

In the stable

Introduce the foal to a companion − this may be another foal or pony − and allow them to spend time and eat together. The mare may then be quietly removed while the foal and his companion are tucking into a tasty meal. The top door will have to be closed, and any potentially dangerous objects re-moved. The foal and his companion should be given many very small feeds and a constant supply of good hay (not in a net) to help provide comfort as well as distraction.

If done quietly and with forethought, weaning need not be too stressful. However, when turning out the foals for the first time try to do so out of earshot of the mares − this is not always easy to organize, as a high-pitched whinny travels a great distance!

Aftercare of the mare

It is important that the mare is turned out onto a fairly bare paddock, since good grass promotes milk production. Her diet, also, must be of a lower quality and quantity to discourage the production of milk.

A little milk may be drawn off each day − do not take too much, as this encourages the production of more. Check that the udder is not becoming hard and uncomfortable − if it does, the vet should be called in case mastitis is developing.

After approximately ten days the mare should have dried up and her diet may gradually be improved.

15

INITIAL TRAINING OF THE YOUNGSTER

It is not within the scope of this book to discuss fully the vast topic of training the horse. However, the handling of the foal, particularly after weaning, leads naturally on to the initial training of the very young horse. This initial training is influenced by the horse's natural instincts, the strongest of which is fear. This fear must be overcome and replaced with confidence and respect — something that can only be achieved through quiet, confident handling.

The foundations of the future work are laid during the early handling of the foal. A summary of this early handling would include:

Quiet acceptance of the foal slip.

Leading in hand and obeying verbal commands such as 'whoa' and 'walk on'.

Standing tied up under supervision.

Having the feet well handled, picked out and trimmed.

Accepting a rug and roller.

Regular introduction to strange objects, sights and sounds to gain confidence and reduce natural fear.

136

The initial training of a riding horse usually begins between the third and fourth year, provided that the horse is well developed both physically and mentally.

One of the dangers of starting the training too soon is that the limbs are not yet sufficiently strong and may be prone to a variety of concussion-related problems. Mentally, the horse may be unable to cope with commands — his short concentration span may lead to confusion and fear.

At the other extreme, it is not necessarily a good idea to leave the training until later (fourth or fifth year). The horse is by then physically strong which, combined with an absence of education, may lead him to be difficult — if not dangerous — to work with.

All horses are individuals — colts tend to be physically stronger than fillies, so may benefit from an earlier start.

THE STAGES OF TRAINING

The initial training may be broken down into the following stages:

Leading in hand.

Introduction to the bit and bridle.

Lungeing, loose schooling and long reining.

Backing — acceptance of the rider's weight.

Riding on lead and lunge.

'Riding away' (early work under saddle).

The amount of time spent starting a horse off — from the first lunge lesson to being safely ridden away — will vary from horse to horse according to temperament. The expertise of the trainer and the facilities available will also affect the time taken. Experienced trainers develop their own training programmes according to personal preference. In an ideal world, all horses would be started off by knowledgeable, patient trainers with as much time taken as is needed to develop a trust and mutual respect. Unfortunately, many horses are hurried through this process by impatient handlers — the longer the time spent on the early work, the better.

These formative months provide the basis for all of the horse's future work — patient handling and firm establishment of the basics will pay dividends later on.

Leading in hand

Ideally this should be done in the early weeks of a foal's life. If a youngster has not yet been taught to lead this must be done in an enclosed space — preferably a very large loosebox. The handler should wear a crash cap and gloves and carry a schooling whip. The horse should wear a lungeing cavesson, to which a long lead rope is attached.

At all times the horse should be encouraged to walk forwards. An assistant can help to send him forwards with a quiet flick of a whip behind the hocks. If an assistant is not available the handler will have to reach back with the outside hand and tap the horse with the whip. The voice must be used in clearly definable tones — this will help when lunge work begins.

Once the horse is used to being led from either side he can be led out of doors — always use a cavesson with a long line for added control.

Introduction to the bit

The type of bit used for mouthing should be a simple snaffle or mouthing bit. A rubber-covered snaffle provides a mild mouth-piece, although the special mouthing bits with 'keys' are preferred by some trainers. A twist of hay around the mouthpiece of a snaffle encourages the horse to 'mouth' at the bit, so helping to keep the mouth moist. Alternatively, a sweet substance smeared onto the mouthpiece can make the horse's introduction to the bit quite pleasant.

When mouthing, the bridle must not be left on an unattended loose horse in the stable as the bit rings or cheekpieces may get hooked up on the top bolt of the stable door. The horse may then panic and pull back, possibly damaging his mouth as well as his confidence.

Lungeing

Educational lungeing of youngsters should only be carried out by persons familiar with lungeing equipment and techniques. Unskilled lungeing will have only negative effects.

Initially, the horse should be lunged wearing only a snaffle bridle with the reins removed, a lungeing cavesson and brushing boots. Attach the lunge line to the centre ring of the cavesson.

The first lunge session should be seen as a natural progression from being led in hand. An assistant remains at the horse's head at his inside shoulder, walking behind the lunge line. The handler encourages the horse to walk out onto a larger circle, keeping him moving forwards through the positioning of the lunge whip. At this stage it is very important that the handler is always 'behind' the horse to send him forwards (level with the horse's hocks, not his shoulders or neck). If the handler is 'in front' of a youngster he then has the opportunity to turn inwards and/or whip round or run backwards.

Work the horse evenly on both reins and keep the circles as large as possible (approximately 20 m) to help prevent excessive strain on the young limbs and muscles. Do not overdo each lungeing session — keep them short, so that the youngster does not lose interest or become physically tired.

The horse should learn to move freely and calmly forward at walk and trot, and to halt.

Many young horses try to turn in constantly — this can be corrected by lungeing from the bit and using two lines. In this case, the roller must be introduced first to allow the outer line (which is attached to the outer bit ring) to be passed through the terret to prevent it from trailing too low. Pressure on the outer lunge line helps to stop the horse from turning in. When halting, ensure that the horse remains on the circle — he should not be encouraged to turn in and walk towards the handler. Again, the outer lunge line helps achieve this.

Introducing the roller

The use of a rug and roller assists the preparation for wearing a roller when working. To introduce a roller in the stable an assistant stands at the horse's head while the handler quietly passes the roller over the horse's back. It may be useful if the horse is allowed to see and sniff the roller first. Initially the buckles are fastened quite loosely and a breastplate is attached to prevent the roller from slipping backwards. Gradually, the buckles are tightened and the horse is walked around the stable to accustom him to the feel of the roller.

Once this has been achieved the horse is lunged or long reined in the roller and, when he is sufficiently relaxed in this, a saddle may be introduced.

Introducing the saddle

Usually an old saddle is used for these early lessons — however, check the condition of the girth straps and stitching, as worn straps can break in the event of the horse putting in a few good bucks. To begin with, a breastplate must be used to ensure that the saddle does not slip back and frighten the horse.

Initially, the stirrup irons and leathers should be removed from the saddle. Wearing the saddle whilst being lunged allows the horse to become accustomed to the sensation and the tightness of the girth. Once the horse is settled, the stirrups may be put back on and fastened up. In due course, they can be allowed to hang down for short spells at walk, to accustom the horse to the feel. However, take care in summer weather that the horse does not try to bite or kick at flies and catch a stirrup iron with his teeth, or a foot.

Introducing side reins

Once the horse has fully accepted the saddle and is working in a relaxed manner on the lunge, side reins may be introduced. The main purpose of the side reins is to teach the horse to accept a contact on the bit whilst moving straight, that is not swinging his hindquarters in or out on the circle.

The side reins must be of equal length and, to begin with, they must be long enough not to have any effect at all. They are attached halfway up the roller or onto the girth straps of the saddle and, prior to use, crossed over the withers and clipped onto the D rings. Once the horse has been trotted on both reins and is calm and relaxed the side reins may be unclipped from the D rings and attached to the bit. (Some trainers prefer initially to clip the side reins onto the cavesson rings before attaching them to the bit.)

As the horse's training progresses, so the side reins may be shortened. However, since the young horse needs to develop his balance and muscles gradually they must never be over-shortened, shortened too quickly, too soon, or used for long periods.

Loose schooling and long reining

Loose schooling and long reining are other methods of working the horse from the ground which can be considered complementary to lungeing.

Loose schooling teaches the horse to respond to the voice commands, whip signals and positioning of the handler, and is therefore considered by many to be a useful precursor to lungeing. It is also a useful alternative, in that the horse can be worked 'large', so there is less stress on young joints than is the case when a horse is permanently on a circle. This, of course, applies only if the horse is worked properly — not if he is allowed to career round out of control. However, carried out correctly, loose schooling establishes a rapport and bond of mutual confidence between horse and handler.

With long reining, the horse is worked from two long lines, one attached to each bit ring. These lines run (in the 'English' method) back through the stirrup irons and round the hind-quarters to the handler, who works the horse from a position about 3 m (10 ft) more or less behind him (usually slightly behind his inside hip). This method is the one normally used for introducing long reining, and for basic schooling. With the 'Danish' method, by which very advanced schooling may be achieved, the reins run through terrets positioned high on a special surcingle, rather than through the stirrup irons.

The advantages of long reining are: the handler has direct contact with both sides of the horse's mouth (just as when using normal reins); there is significant control of the hindquarters; the horse can be worked substantially on straight lines.

A skilful trainer will use a mixture of loose schooling, lungeing and long reining to improve substantially a horse's physique and education. Used incorrectly, or in haphazard fashion, these techniques are counter-productive and potentially injurious to both horse and handler. It is, therefore, worthwhile for anyone involved in the handling of youngstock to seek expert tuition and become properly proficient in these techniques.

Backing

There are no hard-and-fast rules as to when backing may begin — all horses are individuals and will progress at different rates. However, backing should really not begin until the horse has accepted all items of tack and equipment and is going calmly, freely forward on the lunge and/or long reins, accepting the contact. In other words, he must be obedient and confident, and well prepared both mentally and physically.

The process of backing requires the services of an experienced, sensible assistant. In a small, enclosed area the horse should be lunged until calm and relaxed. Remove the side reins and check that the girth is tight enough. The reins should now be attached to the bit. In the early stages the rider can stand on a block beside the horse, level with the saddle, to accustom the horse to a human presence above and behind him. Bear in mind the horse's natural fear of attack from behind — talk to him and reassure him. Only when he is happy with the sensation of someone above him should you try to progress further.

The next step is to apply pressure to the saddle and, when the horse feels ready, the assistant can leg the rider across the saddle to lie on their stomach. This is a very uncomfortable position, and can only be maintained while the horse is led for a few steps. Assistant and rider should continue to reassure the horse. The horse's reaction will determine how quickly to proceed but, after a few sessions, the rider can be gently legged up to sit quietly, without stirrups, leaning slightly forwards. The horse may then be led around and the rider can gradually sit up, keeping a light contact on the reins. Although the handler has the main control over the horse, the rider can begin to apply gentle aids in conjunction with the handler's voice aids. At an appropriate point, the stirrups may be taken.

Once the transitions between walk and halt and turns left and right are established, the lunge line may be paid out so that the horse is being ridden on the lunge. In time, work can be done at trot — initially the rider should do a soft sitting trot, progressing on to rising as the horse's balance and confidence develop.

When the horse is confident and obedient to the rider's aids the lunge line can be removed. The rider should carry a schooling whip to reinforce the leg aids if necessary.

Early work

The early work off the lunge is best done in an enclosed area with a handler on the ground, ready to assist if necessary. As training progresses, the youngster may be hacked out in the company of a quiet, older horse. He can be ridden over different terrain, up and down gentle slopes, through large puddles and over a pole on the ground. Strange objects such as tractors can be introduced — first with the engine turned off, then with it running.

At the end of a schooling session you can teach the horse to stand while you dismount and remount from a block. An assistant should stand at his head and hold your offside stirrup. The assistant can gradually let go of the horse as he learns to stand still.

Canter work must be introduced only once the horse is balanced, calm and confident in trot. If started too early, it can cause the horse to be unbalanced and hurried, resulting in excitement and/or loss of confidence from the horse, thus undoing your previous work.

Whatever the temperament of the horse the early work programme should not be hurried. Throughout the training programme ensure that:

The horse is much praised for good work.

One lesson is established before progressing on to the next.

Sessions are kept short tö prevent boredom and fatigue.

The horse's diet is adjusted according to the work he is doing and his temperament.

This early work provides the foundation for all future training — the unbroken horse, if well handled from birth, should be unspoilt, ready to be taught and to learn. Everything the horse learns now, whether good or bad, will be remembered. Some young horses will always be easier to teach than others; there will always be wilful and difficult youngsters. The art lies in teaching all horses the correct way of behaving, whether in the stable or school, and therefore laying a good foundation for the future — however long it takes.

CONCLUSION

The management of breeding and youngstock will vary according to breed, type and value of the animals involved. For example, less stringent management techniques are applied on some Mountain and Moorland studs in comparison to those in the Thoroughbred breeding industry. Whatever the breed or type, however, a combination of factors contribute to a breeding programme which fulfills the ultimate aim of producing useful riding horses and ponies and/or successful competition animals:

Selective breeding, using only mares and stallions of proven quality and ability, will promote and enhance the positive characteristics which will be passed on to the next generation.

Sound stud management policies are necessary in order to maximize the breeding potential of all stock and to prevent the outbreak or spread of disease.

Good management at parturition and in the first hours and days ensure that a foal has the best start in life. This, combined with skilful handling and correct nutrition, provides the foundation upon which the foal's potential in terms of training and growth can be developed.

If these policies are followed, not only will they contribute greatly towards the production of good quality riding horses and ponies — those which make up the largest proportion of the equine population — they will also increase the chances of producing members of that smaller, more elite group — the equine superstars!

BIBLIOGRAPHY

Allen, W.E. *Fertility and Obstetrics in the Horse.* Blackwell Scientific Publications, 1988.

Davies Morel, M.C.G. *Equine Reproductive Physiology, Breeding and Stud Management.* Farming Press, 1993.

Kramer, L.M.J. and Scott, J.K. *The Cell Concept.* Macmillan Education Ltd., 1979.

Roberts, M.B.V. *Biology — A Functional Approach.* Thomas Nelson and Son Ltd., 1979.

Rose, J. and Pilliner, S. *Practical Stud Management.* BSP Professional Books, 1989.

Rossdale, P. *Horse Breeding.* David & Charles, 1992.

INDEX